Cultivating Science, Harvesting Power

Inside Technology
edited by Wiebe E. Bijker, W. Bernard Carlson, and Trevor Pinch

A list of books in the series appears at the back of the book.

Cultivating Science, Harvesting Power

Science and Industrial Agriculture in California

Christopher R. Henke

The MIT Press
Cambridge, Massachusetts
London, England

For information about special quantity discounts, please email special_sales@ mitpress.mit.edu

This book was set in Stone Serif and Stone Sans by SNP Best-set Typesetter Ltd., Hong Kong. Printed and bound in the United States of America.

Library of Congress Cataloging-in-Publication Data

Henke, Christopher R., 1969–
Cultivating science, harvesting power : science and industrial agriculture in California / by Christopher R. Henke.
  p.cm.—(Inside technology)
Includes bibliographical references and index.
ISBN 978-0-262-08373-7 (hardcover : alk. paper)
1. Agriculture—Research—California—Salinas River Valley 2. Agricultural productivity—California—Salinas River Valley. I. Title.
S541.5.C2H46   2008
630.72′0794—dc22

2008008743

10   9   8   7   6   5   4   3   2   1

To my parents, Doris and Russell Henke

Contents

Acknowledgments

Not a word of this book would have been possible without the support and encouragement of many generous people, and it is my pleasure to thank them.

Although they remain anonymous here, the California farm advisors and growers who acted as my research informants taught me so much about agriculture and agricultural science. I will always be grateful for their openness and for sharing so much of their precious time with me. Although I am at times critical of their work in this book, I have the highest respect for their dedication to producing food and knowledge. I also received a great deal of help with the historical aspects of my research from the California historians Ann Scheuring and Burton Anderson, Bill Barker (now deceased) of the Monterey County Farm Bureau, Axel Borg at UC Davis's Shields Library, Mary Jean Gamble at the Steinbeck Library in Salinas, Mona Gudgel at the Monterey County Historical Society, Norma Kobzina at UC Berkeley's Bioscience and Natural Resources Library, Bill Roberts at the Bancroft Library in Berkeley, and Meg Weldon at the Monterey County Department of Parks.

This project began during my days as a graduate student at the University of California, San Diego, and I was fortunate to have the guidance of wonderful advisors, especially my mentor, Chandra Mukerji. Rick Biernacki, Steve Epstein, Susan Kaiser, Martha Lampland, and Christena Turner also taught me how to think about and research the big questions that drove my interest in this project. So many friends at UCSD also contributed very important feedback (and friendship) that shaped this book, including Mark Jones, Josh Dunsby, Ben Sims, Patrick Carroll, Paul Chamba, Margaret Garber, Theodor Geisel, Tarleton Gillespie, Susan Halebsky, Mark Hineline,

Beth Jennings, Jennifer Jordan, Moses Kärn, Laura Miller, Falk Müller, Bart Simon, and Katie Vann. UCSD's Science Studies Program also provided support for this project through a series of research fellowships.

After graduate school I spent a wonderful and lively year teaching in Cornell University's Department of Science and Technology Studies, where, again, many friends and colleagues had a profound impact on my thinking about science and agriculture. I am especially indebted to Trevor Pinch and Mike Lynch for their support and encouragement of my work, as well as Florian Charvolin, Dimitra Doukas, Leland Glenna, Steve Hilgartner, Bruce Lewenstein, Francois Melard, Cyrus Mody, Karen Oslund, and Elizabeth Toon.

My academic community is now centered at Colgate University, where my colleagues in the Department of Sociology and Anthropology have provided a supportive and encouraging place for me to complete this project and begin new ones. Special thanks go to Rhonda Levine, Mary Moran, and Nancy Ries, who each provided valuable comments on the book manuscript. I must also acknowledge the support and advice of Don Duggan-Haas, Jordy Kerber, Peter Klepeis, Ellen Kraly, Meika Loe, Paul Lopes, Louis Prisock, and Wendy Wall. Colgate also provided support for this project through its Research Council and Division of Social Sciences.

While each of these places where I have lived and worked provided a new source of support for this project, I also have a kind of "placeless" network of colleagues whom I see only once or twice each year at conferences, in the meantime staying in touch through electronic forms of communication. Ironically, these friends have had some of the greatest impact on my ideas for this book, and I would especially like to acknowledge Tom Gieryn and Kelly Moore for being such sharp readers and providing so much useful feedback on this and other projects. I am also grateful to Ben Cohen, Jen Croissant, Jason Delborne, Scott Frickel, Wyatt Galusky, Ed Hackett, Diana Mincyte, Susan Leigh Star, Rick Welsh, and Steve Zehr.

At the MIT Press, Margy Avery, Deborah Cantor–Adams, the editors for the Inside Technology series, and the rest of the production staff did a wonderful job of seeing this book into print. My thanks also to Lara Scott for creating figures 1.1 and 1.2 for the book.

I come from a family of Wisconsin farmers, but I grew up in the suburbs. My grandparents were the last of the farmers in my family, and three of them passed on during the ten years I took to complete this project. I miss them, and this book reflects their influence on my life. This book is dedicated to my parents; now that I am a parent, too, I can fully appreciate their support and love, and I will always be grateful for them. My thanks also to my daughter, Lin, my brother, Ryan, and my parents-in-law, Carl and Teh-Chao Hsu.

After all these debts, there is still one, the most important of all, to my wife, Carolyn. No one has had to read this manuscript's many versions more times than Carolyn, and she has always patiently indulged me with her insight and wisdom. Carolyn, my life and love will always be dedicated to you.

Cultivating Science, Harvesting Power

1 Introduction: Repairing Industrial Agriculture

A Built (Agricultural) Environment

Driving south of San Francisco on California Highway 101, a driver passes through miles and miles of industrial and suburban sprawl, the result of Silicon Valley's explosive development. The scenery remains the same through Palo Alto and San Jose, until the strip malls and housing developments suddenly fall away and are replaced with soft caramel hills, lonely stands of trees, and the occasional cow. Once this transformation takes place, it is not long before the highway bends west toward the Pacific, and roadside signs begin to promote a series of agricultural "capitals of the world," including Gilroy, the garlic capital; Castroville, the artichoke capital; Watsonville, known for strawberries; and Greenfield, the broccoli capital. At the heart of this region is Salinas and the Salinas Valley—the lettuce capital of the world (fig. 1.1).

For the urban traveler racing south toward a vacation in Big Sur or a meeting in Los Angeles, these signs may look like a self-aggrandizing attempt to make something out of nothing. But field after field of crops do, in fact, add up to a massive center of farm production that supplies most of the United States with fresh vegetables. While the Midwest has been called the United States' breadbasket, these California towns and the coastal valleys surrounding them are collectively named the nation's salad bowl. In 2002, California dominated U.S. vegetable production, producing 99 percent of all the nation's artichokes, 58 percent of asparagus, 92 percent of broccoli, 67 percent of carrots, 83 percent of cauliflower, 94 percent of celery, 86 percent of garlic, 76 percent of head lettuce, and 94 percent of processing tomatoes. Despite its image as the home of Hollywood and Disneyland, surfing and sun bathing, California leads the United States in

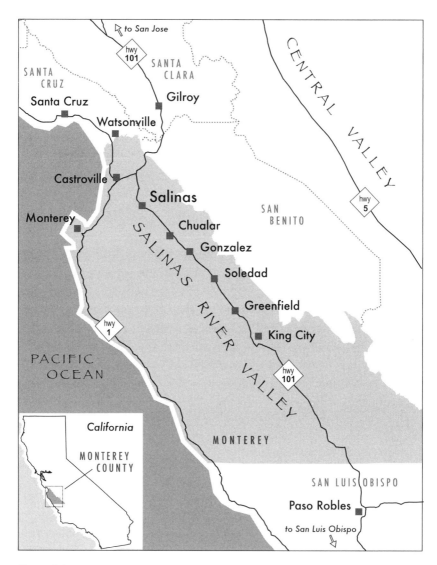

Figure 1.1
California's Salinas Valley. Illustration by Lara Scott.

the overall value of its farm production. At a value of $31.7 billion in 2005, California nearly doubled the cash receipts from farm products in Texas ($16.4 billion), the number two state (CDFA 2006, 20, 25).

How did one small valley become the center of vegetable production for the United States? There are some clues, visible even from the road. The valley's Mediterranean climate allows for the production of many kinds of crops that are not viable on the same scale in other areas of the nation. Large crews of field-workers weed lettuce, pick berries, or cut celery; miles of irrigation pipes and furrows deliver water to thirsty plants; and an endless number of refrigerated semitrucks rumble along, carrying produce destined for salad bowls throughout the United States.[1] Beginning in the summer of 1997, I, too, drove through the Salinas Valley, but I stopped and stayed, in order to satisfy my own curiosity with this question. I spent the next several years exploring an institution that is essential for understanding the history and contemporary context of this industry: agricultural science. During my time studying the vegetable industry, I found that farming in the valley was built on land, water, money, and labor, but the work of scientists has shaped, mediated, and stabilized the relationships between these elements. Agricultural science and farm technologies grew up alongside the farm industry, coproducing each other in very literal ways. Science, therefore, is a relatively hidden but essential element for understanding how the vegetable industry—and California's farm industry more broadly—was created and is maintained on such a broad scale. In short, although the view changes on the drive from San Francisco to the Salinas Valley, the landscapes of urban sprawl and pastoral farmland have more similarities than differences: this valley of vegetables was built, and scientists did an important part of the building.[2]

This book is about the work of agricultural scientists employed by the University of California (UC), and I use the case of UC Cooperative Extension farm advisors in the Salinas Valley to illustrate how scientists and growers[3] have cooperated—and struggled—over how to solve problems associated with building the state's farm industry. These farm advisors are employees of the university but do not work on campus; instead, they are stationed in counties throughout California, charged with providing advice and expertise to their local farming communities. When faced with "crises" as diverse as labor shortages, plagues of insects, and environmental regulations, experts from the UC have stepped forward to help California's growers. Through these interventions, agricultural science has served as a mechanism of repair, a means for maintaining the diverse social and material elements required to grow crops on an industrial scale. Behind the production of a head of lettuce are many seemingly mundane technical decisions like how crops are fertilized or how bugs are controlled, but these same details form the basis for an industry: they make a structure that grows crops but also produces power. Thus, I use the story of how the Salinas Valley became the vegetable capital to understand a more fundamental set of questions about practice and power, and the power relationships among science, industry, and the state.

Industrial Agriculture as an Ecology of Power

Americans have a long tradition of romanticizing agriculture. When we envision a farm or the act of farming, we imagine Farmer Brown on a green tractor, a red barn in the background, among rolling fields of crops and livestock. These images make the terms *farm industry* or *industrial agriculture* seem jarring or even derogatory, but I use them in an analytic sense, to convey both the scale of contemporary agriculture in a place like the Salinas Valley and the social structures that support food production.

The origins of industrial agriculture in California date to the late nineteenth century. With profits from the mining industry waning, California investors looked to farming as a way to make money. These early "growers" farmed massive wheat and barley fields, many of which were more appropriately measured in square miles than in acres (Daniel 1982). However, as competition in grain markets, especially wheat, increased through the late nineteenth century, growers shifted production from field crops to fruit and nut crops. From about 1890 until the onset of World War I, growers experimented with a wide variety of these crops, and the largest producers developed complex production, distribution, and marketing systems for orchard crops like peaches, grapes, raisins, almonds, and oranges.[4]

In turn, these systems served as models for the growth of other farming regions in the state, forming niche market industries, commodity markets that, although relatively small compared to the so-called major crops such as corn or wheat, are still large enough to constitute a significant market for large-scale production.[5] Niche industries are formed around intensive crops such as fruit, nut, and vegetable crops that may rely heavily on hand labor for planting and harvesting, require more intensive use of water and other farming inputs, and are often much more expensive to produce than extensive field crops such as corn, wheat, or cotton. The requirements for intensive crops are offset by the potential for much greater profit. For example, in 2003 an average acre of corn in the United States was worth $334, whereas an acre of head lettuce brought $6,370 (USDA 2004a; 2004b). At the same time, this acre of lettuce cost thousands of dollars to produce, creating a situation where the potential risk and profit for the grower are both relatively high. For this reason, farming in California has often been likened to the fortune-seeking, risk-taking ways of the miners who flocked to the state during the Gold Rush. This characterization has often been used in a pejorative sense, as when Carey McWilliams, in his sweeping work of social criticism, *California: The Great Exception* (1949), wrote, "The soil [in California] is really mined, not farmed" (101). But the capital-intensive character of California's niche market industries does encourage a kind of speculator's logic. This logic was emphasized to me in an interview with the owner of a large fertilizer company in the Salinas Valley, who described the attitude growers in niche markets often take toward growing conditions:

You don't make money in the produce business when everything is right. If the weather is beautiful and you get a good crop, the odds are you're gonna sell it for [little profit]. It's just when something happens—weather events or something like that—that causes a decrease in production and the market goes up.

The ideal for the niche market grower, in this view, is for a catastrophic event to destroy everyone else's crop while leaving the grower's own crop healthy and ready for a seller's market. This "moral economy" of niche market growers places a special emphasis on California growers' interest in controlling their farming environments and helps to explain their attitudes toward and investments in agricultural science.[6] Although agriculture and science may seem like disparate activities, they have some common features and goals. Perhaps the most fundamental of these similarities stems from the unpredictability of farming. Industrial agriculture is a complex built environment, but changes in the weather, shifting pest pressures, and fickle markets all make agriculture an uncertain venture. Farming is organized on a seasonal cycle with the optimistic assumption that conditions will be more or less the same from year to year, but things are almost never the same. For this reason, agriculture has been called "the first empirical science" because it is essentially a seasonal experiment (Busch, Lacy, and Burkhardt 1991). Growers have always experimented with new techniques and technologies to improve yield and quality while at the same time seeking to understand the complex interaction of soil, water, climate, and life to improve conditions of predictability and control.

In this sense, science and agriculture share a practical interest in a kind of mastery of the world, disciplining and systematizing it into a form that reduces but does not quite eliminate uncertainty. In neither case is this mastery a matter of purely academic concern; instead, the ability to effectively control people and things is a critical source of power.[7] Mundane technical practices like how crops are fertilized or how bugs are controlled may not seem closely related to lofty matters like power, but there is an essential link between them. The fertilizer dealer quoted previously emphasized the local conditions for farming, the influence of commodity markets, and how these factors interact and shape each other. To understand the relationships between growers and agricultural scientists, it makes sense to study the multiple levels of material and social stuff that are the subject of their work. The difficulty with this approach lies in theorizing the connections between these diverse factors; scholarly research on science and agriculture tends to focus more strongly on one or another level of analysis. For example, work in science and technology studies (STS) concentrates most often on the local culture and practice of scientific communities, whereas research in the sociology of agriculture emphasizes the political economy of agricultural markets, industry organization, and state institutions.[8] My goal here is to develop analytic tools that bridge these levels of analysis, to understand the ways that local interactions are connected with institutional structures. I use two interrelated concepts to understand these relationships.

First, I conceptualize the diverse social and material elements behind industrial agriculture as an *ecology of power*, a broad system of social and material production that forms the larger playing field where growers and agricultural scientists work to turn products created from local contexts— food, commodities, data, knowledge—into capital that is transferable to other institutions. These forms of economic and social capital are made valuable through this process of exchange and, in turn, can provide actors with control and influence over the very places and practices that serve as the basis for this capital. Despite actors' best attempts to control the production of capital from this ecology, however, there are numerous sources of disruption that challenge their mastery of the structure; anything from a dry year or a failed experiment to a budget crisis or a war can affect these exchanges. These disruptions create the need to *repair* these flows of production—the second conceptual tool that I use to understand the power relations between science and industrial agriculture. Repair work takes diverse practical forms, shaped by the interests of actors in how an ecology produces capital and power. I argue that agricultural science has often served as a mechanism of this kind of repair for the farm industry, working on nearly every aspect of the ecology in order to maintain the productivity and power of the industry.

Figure 1.2 presents the key elements that compose the ecology of power for science and industrial agriculture. My use of the term *ecology* is borrowed from biology, of course, but also from the field of science and technology studies, where the concept of institutional ecology is used to analyze complex social and material networks of activity.[9] The metaphor is useful because an ecology is a field where elements interact in a hierarchical, interdependent, dynamic system. Figure 1.2 schematically represents the interactions of land, plants, scientists, growers, farmworkers, farming techniques, and social institutions that produce not only food but also wealth, knowledge, and, ultimately, power. Changes to one element in this system affect the others; the management and control of each element provides control over the larger system of production and power. In this sense, power is the ultimate product of this system, but power itself is not a tangible quality or "good" apart from the social and material interactions of this ecology. Power is an effect that is produced through this interactive process, and control of the process is the key to power.

There are three levels of analysis in this model (see fig. 1.2, bottom to top). Although the distinctions between these three levels of analysis are artificial, they help to clarify the interactions and influences between different parts of the ecology, especially those elements that are less visible yet essential parts of this system of production. The first level, the local context, forms the base of the model and represents the local *place* where both agriculture and applied agricultural science happen. The unique interactions of soil, water, climate, and other place-specific factors compel both growers and scientists to account for the local characteristics of a given

Figure 1.2
An ecology of power. Illustration by Lara Scott.

place, adapting their work to local conditions.[10] This engagement with the material world also provides actors with skills that can be formally or informally codified in the form of *practice*, the second level of analysis. Some farming practices are quite old: the use of furrows for irrigated agriculture began thousands of years ago and is still used to this day. Whether new or old, however, these ways of interacting with the contingencies of place represent a kind of investment, and changes to even seemingly simple practices can lead to large-scale disruption of the overall production system. In the same way, scientists themselves have considerable interests in practice, which define their research careers and serve as a kind of structure for their work (Pickering 1980; 1984). One of the most common ways for agricultural scientists to convince growers of the value of a new way of farming is through the use of field trials, experiments that use a plot of land to test and visually demonstrate the efficacy of a new practice or technology (see chapter 5).

These local combinations of land and practice are used to create portable capital that is valuable and transferable to the third level: *institutions outside of a given place* (Bourdieu 1990; 2004). Growers produce crops; those crops are sold on commodity markets and transformed into wealth. Similarly, applied agricultural science combines place and practice when testing new farming techniques to assess their efficacy. These experiments create new knowledge that may be translated to other contexts or used to reshape the very relations of practice and place that produced the knowledge. In each case, farm commodities and knowledge are relatively stable products that may be used as capital and exchanged for other forms of capital as well as to control and reshape the overall ecology itself (Latour 1988; 1990; 1993). This control is the basis of power, but it is a "fragile power," and seeing how capital is produced from the most basic interactions of practice and place makes it easier to understand why actors may be intensely interested in the impact of some form of disruption to production.[11] Saying that "knowledge is power" or "wealth is power" is, in this view, inaccurate; knowledge and wealth are only as powerful as the places and practices on which they are based.

The value of an ecological approach to the analysis of industrial agriculture is that it does not emphasize place, practice, technology, markets, politics, or culture over any of the others but instead seeks to understand the interactive effects of these elements—"ecological determinism" is a contradiction in terms. This approach is in contrast to a long history of social science methods that have emphasized (and even naturalized) the effects of economic and technological change in agriculture, placing normative labels on growers and their adoption rates, or conceiving of change as driven by irresistible treadmills (Rogers 1958; 1983; Cochrane 1993). While profit motives and new innovations certainly have powerful effects on the decisions and interests of both growers and scientists, an ecological view of production highlights actors' attempts to control the very factors that create these results. Capital is not self-generating, and in this way, a control motive is essential for understanding a complex system of production like industrial agriculture (Noble 1977; 1984). When the production of something as simple as a head of lettuce or a stalk of celery can lead to power, control of even small details of production is a key interest for the players in this ecology.

Controlling Agricultural Ecologies: A Theory of Repair

If an ecology of power represents the material, practical, and institutional structure of the relationship between science and industrial agriculture, then repair is the work of maintaining this system in the face of constant change. In everyday usage, "repair" describes a process of fixing things. Sociologists working in the fields of ethnomethodology and symbolic interaction often use the term in a slightly different sense: a process of maintaining social order. In this view, social order is a practical, everyday accomplishment, negotiated over and over again.[12] The canonical example of this negotiated order comes from studies of everyday conversations, where social interaction is like a juggling routine between two actors. In the course of a conversation, actors skillfully toss all kinds of "objects" to each other; in this case, the objects are symbols, meant to be understood and returned. In this metaphor, meaning and understanding are contingent upon the continual exchange of symbols and their proper recognition as such. Just as in a juggling routine, a misthrown symbol requires an adjustment to continue the interaction. We continually adjust to slight variations in the flow of symbols as we communicate with each other. Actors in social interaction work within a basic structure of meaning, but the actual accomplishment of this interaction is highly improvisational. In this way, the ethnomethodological sense of repair provides a rich view of social life, where order, adaptation, and change are structured yet fluid, omnipresent yet delicate.[13]

This way of thinking about social order works well for understanding some aspects of the ecology of power that I have described. We can envision agriculture as a kind of structure that is dynamic and constantly shifting. Changes in the context of farming, whether from new pest pressures, volatile markets, or shifting government policies, can threaten the production of the various forms of capital depicted in figure 1.2, requiring growers and scientists to renegotiate and repair their practices to account for these changes. The analogy of everyday conversation as social order, however, only goes so far in explaining the case I am considering here. Prior work has given little attention to either the material context of social interaction or the power dynamics of structures beyond the level of interpersonal interaction.[14] Built environments like industrial agriculture point to the need for a theory of how the social and the material are brought together in systems of production, and how these systems are maintained through the efforts of interested actors. In addition, a broader theory of repair needs to account for disagreements about repair, particularly over how repair should take place or whether something needs to be repaired at all. In disputes over repair, order may likely be the result not of mutual understanding but rather of power relations.

Figure 1.3 presents my framework for a broader theory of repair, including two types of repair strategies and the repair practices used to work toward these strategies. The first distinction, between the repair strategies of maintenance or transformation, points to the different goals and investments that actors may have in the particular structure of their social and material ecology. Does this structure create capital for an actor? Does the structure limit or create barriers for another actor and his or her interests? These factors help explain why a maintenance or transformation strategy of repair would be preferable to a given actor or group. Repair as maintenance is an attempt to solve problems by making modest adjustments to the elements within an established structure, keeping intact as much of the system as possible while remedying the trouble. Repair as transformation is a more radical set of changes to the actual ecology, in which the relationships between culture, practice, and environment are substantially reordered. In most cases, repair as maintenance is the default strategy, especially for those actors with significant investments in the existing

Repair Strategies		
Repair Practices	*Maintenance*	*Transformation*
Discursive	—reinforcing and protecting established meanings and boundaries —framing "problems" as threats to established structures of power —downplaying critiques of established structures	—challenging and calling for change to established meanings and boundaries —framing "problems" as endemic to established structures of power —critiquing and proposing substantial change to established structures
Ecological	—responding to problems through established structures —modest change of established structures in order to preserve overall control over production —creation of new structures that ultimately preserve established systems of production and reproduce power	—responding to problems through structural change —large-scale structural change to established systems of production —creation of new structures that disrupt established systems of production and power

Figure 1.3
Repair strategies and practices.

structure of an ecology. At first, this may seem obvious: it will always be more convenient for actors to maintain a system rather than create a new one. But we only really understand why it seems convenient to maintain a system when we uncover its structure and explore the interests at stake in the balance between order and change. In many cases, maintaining the structure of an ecology may take just as much effort or as many resources as transforming it, but maintenance of the status quo is beneficial for a select group. Therefore, transformative repair is much more likely to be proposed and supported by those who are critical of, disenfranchised from, or subject to an established ecology of power. Those who benefit from the ecology are only likely to turn to transformative repair when the structure no longer works in the same way for them or as a last resort.

A second distinction in figure 1.3 emphasizes the sociomaterial approach I am taking here. The two repair practices—discursive and ecological—describe the practical methods that actors may use to repair an ecology as well as the diverse forms that this improvisational work may take. Discursive repair is the form that has been studied most extensively through ethnomethodological conversation analysis, but the focus I take here is on a broader view of discourse, where repair is aimed at the cultural and symbolic frames that shape and legitimate structures.[15] This repair practice is a kind of "boundary work" (Gieryn 1983; 1995; 1999), a way to discursively maintain or transform the frames around an ecology, but it can also cast repair itself as the subject of debate and negotiation. Does a problem exist? What exactly is the nature of that problem? And what is the best way of solving it? The answers to these questions represent practical attempts to shape the discursive frame for meaning and action, which, in turn, lead to ideas about what form a structure can or should take. In contrast, ecological repair is aimed at the institutionalized practices and material structures that shape the production of capital within an ecology. Clearly, these two sets of practices overlap, and, in fact, the case studies that I present in subsequent chapters show that most repair draws on both. The work of rhetorically defining a problem for repair goes hand in hand with institutional forms of repair, providing a continuum from methods of insect control to large-scale legitimation crises of the entire ecology of power.[16] The following section describes how growers and scientists have historically navigated this ecology of power and negotiated its shape through repair.

Science, Industry, and the State: A Brief History of Repair in Agriculture

The ties among agriculture, knowledge, and power have long been understood by the state; without an adequate and affordable supply of food, it is difficult to maintain rule for very long. Archeological evidence suggests a relatively direct link between the stability of food production and the viability of a political elite (Diamond 2005). This relationship is also expressed in the stories we tell about food production and politics. For example, in the Bible's Genesis story, Joseph, sold into slavery in Egypt by his jealous brothers, found favor with Pharaoh by interpreting a set of strange dreams as a prophecy about seven years of bountiful harvests followed by seven years of famine. Pharaoh was so impressed by this advance warning that he put Joseph in charge of a project to hold back grain during the plentiful years, in preparation for the lean ones. Early in U.S. history, Thomas Jefferson believed that the nation's democracy was directly dependent on the strength of its farming and saw the independent family farm as a stabilizing force against class conflict and economic change.[17] Even now, when less than 2 percent of the U.S. workforce is in farming, compared with nearly 50 percent at the beginning of the twentieth century, many lament the demise of the small family farm and the public good it is presumed to provide.

Beginning in the latter half of the nineteenth century and continuing through today, the state and other farming interests have sought to control and protect U.S. agriculture through the use of science. Long before the work of physicists and engineers was seen as essential to national security, agricultural science was proposed and accepted as a state project, intended as an antidote to foreign competition, increasing cost-of-living expenses, and a perceived decline in the quality of American rural life (Rosenberg 1976; 1977; Marcus 1985). The first major step toward a formal system of state-sponsored agricultural science began with the foundation of the land-grant university system following the Morrill Land-Grant College Act of 1862. This federal legislation provided grants of land to states on the condition that they would be sold and the profits used to found a college for training rural citizens in agriculture and other practical crafts, a mandate that came to be known as the land-grant mission. A system of agricultural experiment stations, based on the land-grant university campuses, was also begun in order to provide research on problems important to farming.

Through the first several decades of the land-grant system, however, scientists and farm communities remained relatively isolated: agricultural researchers did not necessarily want to work on problems of immediate practical interest to growers, and growers were often impatient with the promises of long-term basic research (Marcus 1985). Therefore, around the beginning of the twentieth century, a movement formed that called for a system of extension work to be put in place, to bring research from the land-grant schools to local farm communities. Members of the extension movement argued that a system of extension advisors stationed in local farming communities could bring improved methods of agriculture to rural populations, thereby fulfilling the land-grant mission. In 1914 the federal Smith-Lever Act was passed, providing funds for each land-grant university to establish its own system of extension work. The program was named Cooperative Extension because funding for the advisors came from the federal, state, and county levels of government. After a rapid expansion of the program during World War I, most counties in the United States had a Cooperative Extension advisor, affiliated with each state's land-grant university and charged with promoting the latest methods and technologies for making local agriculture more productive. At that time, and likely still today, Cooperative Extension represented the most widespread and pervasive arm of state-based expertise in the United States.

Cooperative Extension work is an ideal case for studying the interface of social and material repair because advisors are supposed to intervene directly in the ecology of place, practice, and power found in their local communities. This aspect of Cooperative Extension work brings to mind Michel Foucault's extensive work on the ties between power, knowledge, and the modern state. In Foucault's conceptualization, the state maintains power and social order through modern institutions of expert knowledge and practice, such as medical clinics and prisons. In this respect, Foucault's view of modern statecraft is defined not so much by an ideology or a historical era but rather by institutionalized practices, a set of techniques that consolidate power.[18] Despite Foucault's emphasis on the power of these practices, however, their efficacy seems quite variable. The techniques of domination he describes in connection with the rise of penology have had a profound effect on the way modern societies treat lawbreakers, but new problems seem to inevitably arise from these systems of power-based order.

This dichotomy between state intentions and outcomes is the subject of James Scott's *Seeing Like a State* (1998), in which he describes several "high-modernist" attempts by the state to control and direct the lives of its populace, including urban planning, agrarian reform, and rural resettlement projects. In each of his examples, Scott writes, the state viewed this intervention in a very linear and simplistic way, assuming that abstract principles of design imposed from above could easily improve and replace the systems of practice already in place on the local level. In addition, state planners believed this reordering of local practices could increase the control and accountability that the state held over its people, "consolidating the power of central institutions and diminishing the autonomy of [subjects] and their communities vis-à-vis those institutions" (286). Scott describes how these grandiose projects turned out quite badly for the state and especially for its people, and he argues that high-modernist projects failed because they did not account for the importance of local knowledge. Overall, the cases presented by Scott argue against the totalizing power of the modern state and its ability, as emphasized in Foucault's work, to intervene in local communities of practice (1998, 101).

The case of Cooperative Extension lies somewhere between these two extremes, where expert-based power is portrayed as either totalizing or bumbling in its control of the local. Unlike the projects described by Scott, Cooperative Extension was intended to intervene on the local level by specifically accounting for the importance of place and practice. Although it has meant different things to different people, Cooperative Extension was designed as a way of influencing farm practices by putting a network of experts directly in contact with farm communities. Given this difference, farm advisors' work at first seems more in line with the kind of knowledge-based techniques of mastery that Foucault describes. But when the state does intervene in a decentralized way, the results are mixed, and do not always resemble the kind of totalizing power that Foucault attributes to the rise of expert systems of control (Mukerji 1997, 321). As a result, my analysis here is aimed in a slightly different direction, toward understanding how these experts themselves—farm advisors in this case—become entangled in existing and ongoing power struggles that tie local places and practices to the state and the wider farm industry.[19] As depicted in figure 1.2, the knowledge and expertise of advisors (and the university) has influence on other state actors and the farm industry, but these same groups also shape the work of advisors through their own sources of influence, including wealth and political regulation. In the face of constant change and crisis from many sources, Cooperative Extension has served to stabilize and reproduce the ecology of power in industrial agriculture, repairing elements on every level of this ecology. At the same time, the story of Cooperative Extension is just as much about how local actors resist and reshape state institutions. As one example, during a series of labor crises brought on by the onset of World War II, the farm industry demanded that Cooperative Extension advisors organize and ration the use of diverse sources of farm labor, essentially acting as a kind of labor contractor (see chapter 4). Farm advisors cross many boundaries in the course of their work, but the common thread among this work is repair; Cooperative Extension is an institution of repair. Its mandate to improve the productivity of agricultural communities has often served to preserve and maintain the power structure of the local social and material ecology.

The Structure of This Book

Science and industrial agriculture are engaged in a long process of historical and cultural coproduction. Because I want to understand and see this process at multiple levels of analysis, I used a combination of methods for this project, including participant-observation fieldwork, semistructured interviews, and analysis of historical documents. My choice of these methods for a study of repair was inspired by Jean Lave's call for a "more inclusive theory of social order" that sees "objects of analysis [as] points of cultural-historical conjuncture [that] should be analyzed in those terms" (1988, 171). Therefore, I see a multimethod approach as essential for a "thick" ethnographic analysis of the relationship between industrial farming and agricultural science (Geertz 1973). See the appendix for details on my data collection methods.

My analysis of Cooperative Extension as an institution of repair draws on both historical and contemporary examples, but the overall structure of the cases is largely chronological. Chapters 2 and 3 each cover the history of Cooperative Extension in the context of the United States, California, and the Salinas Valley. In chapter 2, I describe the history of the agrarian ideal in U.S. agriculture and especially how that ideal changed in the context of industrialization and social change during the nineteenth century. These shifts created a cultural context in which agriculture could be seen as a social problem and in need of repair, leading to the establishment of the land-grant university system. In particular, I trace the ideals and interests of Progressive Era politics, circa 1890–1920, and social movements that called for the use of expert knowledge to address the perceived deficiencies of U.S. farming and rural life. Though Cooperative Extension's mission of service to local farm communities derived from this ethos of expertise, the practical implementation of farm advising was subject to divergent interests and a great deal of uncertainty about the best direction for U.S. agriculture. Using archival materials and oral histories, I trace the formation of UC Cooperative Extension and describe the growing pains it faced when confronted with the rise of niche industry farming in California in the period between the world wars. Chapter 3 continues this story for the case of farm advising and the produce industry in the Salinas Valley. Cooperative Extension's "mission ambiguity" was at least partly resolved for the Salinas Valley farm advisors with the rise of the vegetable industry—an industry hungry for specific technical expertise and armed with the funds to support its production. At the same time, this specialization continued to raise questions about the appropriate relationship between advisors, industrial agriculture, and smaller growers.

Chapter 4 treats the issue of farm labor during World War II and details how growers and advisors tried to respond to a labor crisis during the war years. Although labor was an important problem for growers during this period, I show that the war years were actually part of a long series of labor "crises" in California's farm fields. Both within the farm industry and among experts and intellectuals, this recurring history of labor conflict represented an irrational problem that merited a rational solution, but the definitions of the problem and the proposed solutions were diverse. With these ongoing debates as a larger context, I use the case of the Spreckels Sugar Company to show how the farm industry and the UC worked to resolve the farm labor problem in the years before, during, and after World War II. The case makes a useful example of how labor, technologies, grower practices, experimental knowledge, and the production of commodities are enmeshed in a struggle to maintain control over the larger ecology of power in industrial agriculture.

In chapters 5 and 6, I focus on more contemporary instances of advisor-grower interaction. Chapter 5 describes how farm advisors use field trials to collect data on new farm practices and to convince growers that these new practices are worth adopting. These trials make a useful case for exploring the role of place in the negotiation over what set of practices makes up the best way to farm a given piece of land. Typically, growers are most likely to trust research results that are generated from a trustworthy place—often their own land—and so field trials conducted on a grower's own field can make a powerful demonstration. At the same time, the place-bound character of these trials raises many issues of control for the advisors and their experimental practices. Therefore, field trials are an excellent site for studying in detail how place and practice shape, and are shaped through, the use of experiment.

Chapter 6 also focuses on the balance that advisors strive to maintain between their ideals for new farming practices and the investments that growers may have in existing methods. In this case, the balance concerns environmental problems related to farming in the Salinas Valley. Today's farm advisors spend a significant amount of their time working on environmental issues, trying to minimize the environmental impacts of farming in the county. Growers also see these issues as problems, but not in the same way; they may be primarily interested in deflecting criticism of their industry and preventing further government regulation. These alternative definitions of the situation can lead to conflict, and I analyze advisors' work on environmental problems as a special case of boundary work, where advisors walk a fine line between an established order of practice and the possibility of social change (Gieryn 1983; 1995; 1999). Overall, chapters 5 and 6 provide a detailed look at how repair works at the local level and how this local work is integral to the maintenance of power.

Chapter 7 draws upon the preceding empirical chapters to further elaborate my theoretical interest in the relationship between different kinds of production, especially the interface of commodities, knowledge, and power. I argue that the study of repair is broadly applicable to a range of other cases, especially where the maintenance of power relations is a key concern for actors. In addition, I explore the policy implications of this work and end the chapter with a vision for how expertise could be deployed to encourage more extensive change. I argue that farm advisors have an advantageous position as locally based sources of expertise, but that they are also limited by this position. Advisors need additional support from the state or from other sources of leverage that can make their expertise more widely available and effective at promoting change. Though additional (direct) governmental regulation is one way to give advisors more power to influence change, I suggest that ethnographic accounts like mine also play an important structural role in the process of social change.

2 A New Agrarian Ideal: Foundations of Cooperative Extension

What Needed Repair, and Why?

If Cooperative Extension was created as an institution of repair, what was it supposed to fix, and why? The answer starts with the profound changes that began to transform U.S. agriculture during the nineteenth century. Demographics do not tell the entire story, but consider that about 80 percent of working Americans were farmers in 1800, whereas fewer than 40 percent were by 1900 (Cochrane 1993; Dimitri, Effland, and Conklin 2005). As Frank Norris described these changes in his novel *The Octopus* (1901), the rapid pace of industrialization entangled U.S. agriculture in powerful effects from the rise of commodity markets, increasing urbanization, and the diminishing effects of time and distance through new technologies such as the railroad and the telegraph. While agriculture became more tightly entwined in the institutions of industrial modernity, national views about farming and rural life began to shift. More specifically, beginning in the late nineteenth century, Progressive Era thought created a political and cultural context in which agriculture could be seen as a kind of social problem, in need of repair through state-sponsored expertise. This view of farming turned the classic Jeffersonian vision of agriculture—where the independence of the small family farm served as the foundation of U.S. democracy—on its head. These changes set the context for a new vision of agriculture and a role for agricultural science in creating it.

Overall, the cultural, economic, and political status of farming in the United States at the turn of the twentieth century stood in stark contrast to its position 100 years earlier. Cooperative Extension was created in the early decades of the twentieth century in order to bring U.S. agriculture into the industrial age. But once Cooperative Extension had been established, a key question remained: what exactly should Cooperative Extension do to make U.S. farming better? In this chapter I describe and analyze the cultural context for extension work in its early years. The overarching frame for the debate over the status of U.S. agriculture was a dichotomy between the industrialization of farming and the classic agrarian ideal of small independent farms. These contrasting ideals shaped the debate over what was wrong with agriculture and how it ought to be fixed. In addition, key ideas in Progressive thought—especially the value placed on expert knowledge and efficient methods of working and living— provided an intellectual and political context in which the state could seek to transform farming communities by sending agents into each county of the nation to intervene in a practical and local fashion. Once in place, however, Cooperative Extension advisors found that expertise and efficiency did not change agriculture for the better in and of themselves, and that growers did not necessarily appraise these values in the same ways.

The debates about how to improve agriculture were not resolved by Cooperative Extension; instead, these competing discourses formed the basis of a kind of institutional identity crisis for farm advisors. Cooperative Extension was born at the same time that the California farm industry was beginning to take shape, and so its identity and mission were tied up in the same questions about what was the best kind of agriculture for the state and nation. Farm advisors had new techniques and technologies to offer growers, but advisors' expertise was not necessarily useful for all growers in the same ways. Therefore, the historical development of Cooperative Extension makes a good case for exploring situations where repair itself is a contested category and actors negotiate the techniques and extent of repair. Understanding this complexity is important for understanding the structure of Cooperative Extension and the work of advisors. Cooperative Extension was created as an institution to promote transformational change in U.S. agriculture and rural life, but advisors could not simply wipe existing practices, structures, and power relations of agriculture away. Their work began in a context of industrialization that had already been reshaping agriculture in California for several decades. In this way, the work of farm advisors helped to create the ecology of industrial agriculture in California, but this nascent ecology of power also shaped Cooperative Extension.

The Land-Grant Mission and a New Agrarian Ideal

Ideals about farming and the place of agriculture run deep within the politics and culture of the United States. Perhaps no idea is more important in this discourse than the ideal of the small family farm. Jefferson envisioned a strong rural farming population as the foundation of U.S. democracy: small, self-sufficient agriculturalists would prevent marked class distinctions and protect the economic stability of the nascent democratic system.[1] Thus, the small farmer became a cultural icon, a bellwether of U.S. moral and political sensibility. In the second half of the nineteenth century, however, attitudes toward farming and rural life began to change. While Jeffersonian sentiments glorified rural life and the hard-working, independent agriculturalist, this new view portrayed farm life as a poor, degraded existence. Many began to speak of a "farm problem" and consider ways to elevate the "mental culture" of farming communities (Marcus 1985, ch. 1).

This rhetoric of farming's decline must be placed within a context of economic and cultural change that took place in the United States during the nineteenth century. William Cronon's (1991) work on the rise of Chicago shows how the development of railroads and the expansion of commodity markets increasingly tied growers to financial interests in this period. These connections meant that growers became more vulnerable to shifts in the economy; downturns in commodity prices, especially in the last quarter of the nineteenth century, drove this dependence home. Many in the countryside saw urban financial interests as a parasite on the productivity of U.S. agriculture and cities as seductive magnets for rural youths. Farmers acknowledged their fading sense of independence by forming farm associations such as the Patrons of Husbandry (better known as the Grange) and the Farmers' Alliance (the Populists). These associations became strong political players on both the federal and local levels, mitigating the vulnerable status of agriculturalists but also leading to conflicts with industrial and financial interests. In addition, this period also saw increasing power and protest from urban workers and their unions. The rise of unionism contributed to urban politicians' and capitalists' fears that agriculture and rural life in the United States lagged behind the development of nonagricultural industrial production and urban standards of living. With decreasing productivity in rural agriculture would come increasing costs of living in urban areas, especially troubling for employers in times of labor unrest and increasing unionism. These interconnected factors—the rise of farm-based political groups (especially the Grange) and the increasing sense that farm productivity and efficiency were inferior to other industries—shaped the sense that something was wrong with U.S. agriculture, and commentators began to suggest ways in which it might be repaired.[2]

This changed perception still regarded the health and vitality of U.S. farming as a crucial foundation of the nation but eliminated the ideal of independence from the Jeffersonian vision and replaced it with plans for intervention and repair (Danbom 1979). For instance, some critics suggested that farmers had become too focused on physical labor without cultivating the knowledge for a productive and sustainable agricultural economy. Discussions about potential repair solutions also set in place a set of oppositions that actors could use to frame their perception of the farm problem, including knowledge versus ignorance and efficiency versus stagnation. In each case, reform-minded critics of agriculture's decline supposed that supplying whatever was missing from U.S. farming, such as knowledge or efficiency, would restore the health of farming and farm communities, preserving their status as the backbone of democracy.

These assumptions influenced the development of the land-grant system of agricultural research and education in the United States.[3] The land-grant system was founded in the latter half of the nineteenth century, following the Morrill Land-Grant College Act of 1862. The Morrill Act granted states 30,000 acres of land per member of Congress, and the land was to be sold in order to fund universities that would train the rural populace in agriculture and other practical vocations. Many states readily established universities under the land-grant system, including the University of California campus at Berkeley, founded in 1868. In fact, today, when someone refers to a state university in the United States, they are likely referring to a university that was founded under the Morrill Act.

Although the Morrill Act originated the widespread system of state universities in the United States, its legacy extends beyond the campuses. Through its prescription for a certain kind of education for a certain group of people, the legislation created what became known as the land-grant mission. The mission described in the Morrill Act dictates that the land-grant universities focus on "such branches of learning as are related to agriculture and the mechanic arts, in such manner as the legislatures of the states may respectively prescribe, in order to promote the liberal and practical education of the industrial classes in the several pursuits and professions in life." Note that this passage is precise in some senses and quite vague in others. Although ultimate authority about how to implement the Act's mandates is left to the states, the mandates themselves are left "deliberately nebulous and open to varying interpretations" (Marcus 1985, 129). Consequently, the Morrill Act and the mission it decreed for the land-grant universities stood as a rhetorical toolkit for a wide array of actors, each with an agenda for U.S. agriculture, and these actors often had divergent ideas of how the land-grant mission should be approached. For any question one might ask about the land-grant system and its mission, the answers were tied with conflicting interests.

For instance, for the agricultural scientists, the mission was (and often still is) about using research to make new advances in the basic scientific understanding of natural processes that will benefit applied agricultural practices. In the period in which the land-grant schools were developed, the budding profession of agricultural science modeled itself quite explicitly on the example of German agricultural chemistry. Many young U.S. scientists trained there and were indoctrinated with the German university system's emphasis on basic science and experimentation. These researchers, fresh from their formative years in German chemistry departments, argued for classes in the sciences and lobbied for the institutionalization of agricultural research. Many of the administrators who created the individual land-grant schools and shaped their curricula also had these same biases, preferring courses in the classical university tradition over those they considered vocational (Rossiter 1975; Rosenberg 1976; 1977). Thus, the land-grant mission, according to these administrators and agricultural researchers, was to bring the power of rational scientific methods and experimentation to bear on the problems of U.S. agriculture. Science could bring increased productivity, which would, as Rosenberg reports, renew the moral and public good of U.S. farming: "[Agricultural scientists'] ideological stance rested on an unquestioned faith in the transcendent virtue of productivity; to increase the productivity of the soil—to make two blades of grass flourish where one had before—was to act in an unambiguously moral fashion" (1977, 403). This is a crucial point: for agricultural scientists there was little contradiction between increased productivity and the more general health and morality of rural life. Where "two blades of grass flourish," many believed, farm communities would flourish as well.

For many growers, especially those active in the Grange and other farm movements of the late nineteenth century, the land-grant mission was less about basic research and classical subjects than about practical solutions that could be applied to the immediate problems of local agriculture: the immediate conditions of the soil, the crop varieties to be planted, and other specific details about local farming requirements. Grower-oriented political movements like the Grange and the Populists called for university researchers to concentrate on the practical aspects of farm education, pre-ferring that land-grant institutions teach vocational farming courses instead of the classical courses that private universities traditionally taught. Further, many smaller growers were suspicious of "book farming" and of intellec-tuals who claimed that their new knowledge and advice could increase growers' yields. This sounds like a stereotype of growers as coarse anti-intellectuals, but, in fact, growers were often exposed to and were wary of "magic" formulas hawked by salesmen with claims of increased yields. In addition, growers understood better than anyone the importance of local conditions and their interaction with farm practices. If a new way of farming did not mesh well with the local conditions of land, climate, and farming practices already in place, then research results were useless, no matter how promising they seemed on the university campus.[4] Larger and wealthier growers, by contrast, were often eager to accept the new methods and technologies developed by the land-grant institutions. In fact, it was many of these wealthier growers who promoted the creation of a system of land-grant-based experiment stations under the Hatch Act of 1887 (Marcus 1985).

As an example of these conflicting goals and how they played out in the context of the early land-grant system, consider the early history of California's land-grant institution in the 1870s. The University of California shared the interventionist mission of the Morrill Act and was subject to the mission's ambiguity from its very inception. For example, the university's first professor of agriculture, Ezra Carr, believed, as many in the Grange and Populist movements did, that practical, hands-on train-ing for farmers was the primary responsibility of the land-grant school (Scheuring 1995, 14–15). Taking a different view were those, like Daniel Gilman, first president of the UC, who believed that the university's mission was to build a "foundation for the promotion and diffusion of knowledge—a group of agencies organized to advance the arts and science of every sort, and to train young men as scholars for all the intellectual callings of life" (Scheuring 1995, 15). Despite strong support from the California Grange, Carr's vision for the College of Agriculture lost out, and Carr was dismissed in 1873, just four years after he took the job. Eugene Hilgard, German-born and educated in the German university system, was Carr's replacement. Unlike Carr, Hilgard believed in the primacy of scientific research and education, claiming that the university's mission should be "aimed at progress for the industry rather than training in specific farm skills."[5]

Thus, opinions about the land-grant mission and its implementation were divided from the start. Even when different actors and interest groups agreed that a farm problem existed, their prescriptions were rarely the same. But the sense remained that something needed to be done to improve farming in the United States. Increasingly, farm life was seen as backward and out of step with the lifestyle and efficiency of modern urban life and industry. As the nineteenth century ended and the land-grant system approached its fortieth anniversary, this perception increased and was combined with the philosophical and practical concerns of Progressivism.

Progressivism and the Country Life Movement: Foundations of Cooperative Extension

The Progressive Era, roughly from 1890 to 1920, followed the widespread changes and conflicts of the late nineteenth century. Historical work on Progressivism is often divided on key questions, but certain elements of Progressive thought stand out as generally accepted tenets.[6] These are especially important for the study of Cooperative Extension in California agriculture, for several reasons. First, Progressive thought and politics had a large impact on California in the early decades of the twentieth century, in large part as a reaction to the control that "machine politics" held over the state. The Southern Pacific Railroad held a monopoly over the state's railways, elections, and courthouses, and machine-style political systems in both San Francisco and Los Angeles controlled the policies and purse strings of California's two largest cities. The election of Progressive Governor Hiram Johnson in 1910 was a turning point in breaking this control

(Olin 1968; Starr 1985). Second, Cooperative Extension's genesis was inspired in part by a Progressive movement intended to improve the living and working conditions of rural farm families: the Country Life movement. Studying Progressive thought and its influence on the Country Life movement helps to explain the rationale for state-based intervention exemplified by Cooperative Extension. Third, a key concern of Progressive thought—the social consequences associated with industrialization—was a central, controversial issue in California agriculture.

Perhaps the most important feature of Progressivism was the belief in progress and the power of rational planning by qualified experts to guide this positive growth. Danbom terms this perspective "scientific progressivism" and describes the Progressives' emphasis on the "efficient society":

> Though they were not always clear on specifics, scientific progressives agreed on the general shape of the efficient society. The efficient society, like the efficient factory, would be composed of happy, productive people who would cooperate with one another harmoniously and would accept orders from and defer to an educated elite. This elite would be composed of experts who would make social policy in the interest of the public. Efficiency, at once the means to and the end of the model society, would be the guiding principle for the policies made. (1987, 120)

Given this set of beliefs, Progressivism is often tagged a modernist movement, allied with larger modernist concerns about education, individual reason, and the proper foundation of morality in modern society. Danbom, however, also points to another set of concerns that question Progressives' commitment to modernity: their mixed feelings toward industrialization. Progressives, he writes, looked backward to middle- and upper-class Victorian morals and sensibilities as they also looked forward to a future of increasingly industrialized social patterns and concomitant social problems. In this respect, Progressivism was as much about redemption and repair as it was about progress (Danbom 1987, chs. 1, 2; Bowers 1974, ch. 3). Progressives wanted progress, but it had to be an orderly form of progress. Although Progressives used a discursive frame of "knowledge versus ignorance" to argue for reform, they believed that knowledge was not an easy thing to apply and that experts were required to carry out this intellectual and moral improvement. Therefore, scientific progressivism emphasized the power of state-based planning and intervention to facilitate and oversee this progress.

Progressive experts needed a base from which to act and influence their intended audience, so that groups who needed education and guidance could learn from these authorities and, in a sense, become converted to the Progressives' preferred standard of morality and behavior. This base within the state also points to another key feature of Progressive thought: a distaste for class politics and conflict. Progressives were not the only group fighting for change in a rapidly industrializing nation. Labor unions were also striving for change, and the early years of the Progressive movement, the 1890s, were a period of economic depression in the United States, filled with a great deal of class-based antagonism between capitalists and unions. Progressives aspired to rise above this conflict, and basing expert intervention within the state provided a convenient middle ground from which to intervene in the myriad social problems created by industrialization. Despite this stance of class disinterest, Progressives came mainly from middle- and upper-class backgrounds, and though their social policies and programs were often meant to alleviate the effects of poverty and poor working conditions for the working class, Progressives' concerns were often more closely aligned with the interests of corporate America.[7] Thus, while Progressivism is often considered a reform movement, Progressives' commitment to transformative change was suspect, and there are many indications that their proposed reforms were based on a desire to instead maintain longstanding social and economic boundaries.

These basic characteristics of Progressive thought, including some internal contradictions, are evident in the Country Life movement, an effort by mostly urban politicians, business interests, and agricultural educators to improve the quality of country life in the United States during the first two decades of the twentieth century. The Country Life movement was closely allied with Progressive ideals and policy agendas of the time, but it also drew from the long-simmering sense that U.S. agriculture was not all it could be. Country Life advocates stressed the importance of healthy farm communities and struck a very progressive, interventionist tone about possible solutions. In addition, the urban business interests represented in the movement were concerned about U.S. agriculture's lagging production and possible cost-of-living increases associated with higher food prices (Bowers 1974; Danbom 1979).

The best example of this blend of goals comes from the report of the Country Life Commission, a fact-finding committee organized by President Theodore Roosevelt in 1907. The commission, formed of planners, urban business interests, and agricultural scientists, traveled throughout the agricultural areas of the United States and investigated the status of farm life, reporting back to Roosevelt in 1909. The report reveals how Progressives' belief in "order and rationality in the search for solutions to society's problems," carefully administered by state experts, combined with concerns about agricultural productivity to influence the formation of Cooperative Extension (Daniel 1982, 94). The report begins with an introduction by Roosevelt, explaining his rationale for creating the Country Life Commission (USCCL 1911). In this excerpt, note how Roosevelt blends the concerns I have described:

The Commission was appointed because the time has come when it is vital to the welfare of the country seriously to consider the problems of farm life. So far the farmer has not received the attention that the city worker has received and has not been able to express himself as the city worker has done. The problems of farm life have received very little consideration and the result has been bad for those who dwell in the open country, and therefore bad for the whole nation. We were founded as a nation of farmers, and in spite of the great growth of our industrial life it still remains true that our whole system rests upon the farm, that the welfare of the whole community depends upon the welfare of the farmer. The strengthening of country life is the strengthening of the whole nation.

  If country life is to become all that it should be, if the career of the farmer is to rank with any other career in the country as a dignified and desirable way of earning a living, the farmer must take advantage of all that agricultural knowledge has to offer, and also of all that has raised the standard of living and of intelligence in other callings. We who are interested in this movement desire to take counsel with the farmer, as his fellow citizens, so as to see whether the nation cannot aid in this matter. (9–10)

Roosevelt begins by citing the classic tie between farm life and the nation's life in general, but his words, "The strengthening of country life is the strengthening of the whole nation," provide a sense that this historic bond had faltered and that farmers needed to join urban industries in the process of modernization. On the other hand, by taking advantage of "all that agricultural knowledge has to offer," Roosevelt argues, farmers could raise their standard of living and at the same time develop a more productive food system for the nation. In this view, like that of agricultural scientists, there was no incompatibility between increased production and efficiency and the health of farm communities—indeed, the former could beget the latter.

The text of the report continues in this vein, beginning with laudatory statements about the importance of country life. The commission then addresses the important problems facing U.S. farmers, foremost of which was agriculture's lagging productivity and efficiency: "Agriculture is not commercially as profitable as it is entitled to be for the labor and energy that the farmer expends and the risks that he assumes, and . . . the social conditions in the open country are far short of their possibilities" (18). The report then gives a number of reasons for this "deficient" character of country life, the first of which is "a lack of knowledge on the part of farmers of the exact agricultural conditions and possibilities in their regions" (19). Thus, in good Progressive fashion, one of the commission's first recommendations was to create a system of extension work through the network of agricultural research and teaching resources already in place at the land-grant universities:

We suggest the establishment of a nation-wide extension work. The first original work of the agricultural branches of the land-grant colleges was academic in the old sense; later there was added the great field of experiment and research; there now should be added the third coordinate branch, comprising extension work, without which no college of agriculture can adequately serve its state. It is to the extension department of these colleges, if properly conducted, that we must now look for the most effective rousing of the people on the land. (127)

By bringing knowledge to U.S. farm communities, extension work could increase production and efficiency, restore the public good in agriculture, and make rural life better for all in the countryside.

Although the concept of Cooperative Extension and the use of local agricultural advisors for improving farm practices was not a new idea, the Country Life Commission's influence helped make the system a reality.[8] With the support of the Country Life Commission and other extension movement interest groups, Cooperative Extension was officially created through the federal Smith-Lever Act in 1914. This legislation provided federal money for the states' land-grant schools to fund farm advisors of their own. In addition, the state and county governments that received advisors were to contribute their own funds—thus the name Cooperative Extension, referring to cooperative funding by different levels of government. In many cases, individual states' land-grant institutions had already created their own, more modest extension systems. Cornell University, for instance, was very active in extension work, forming its own county-based network of advisors as early as 1912, due in part to the support of Liberty Hyde Bailey, dean of Cornell's College of Agriculture and an influential member of the Country Life Commission. Many states throughout the Midwest also adopted their own systems of extension work as it became increasingly apparent that federal legislation would be passed to support extension through the land-grant schools (R. V. Scott 1970, ch. 10).

Key to the extension model was the creation not only of a new system of expertise through the farm advisors but also the organization of the farmers themselves. One of the farm advisor's first duties when he[9] arrived in his county for work was to organize a local farm bureau, composed of local growers who were interested in learning about new farming techniques. These farm bureau centers were intended to extend and consolidate the advisors' influence in their county, making it faster and easier to implement new practices. By meeting many growers in one gathering, advisors could multiply their effectiveness. In addition, the farm bureau centers were also supposed to be a place for rural people to meet with their neighbors, to discuss important issues and to be entertained.[10] These same functions could likely have been performed through other, already established farm organizations, especially the Grange, which also had chapters in counties throughout the United States. The Grange, however, was an outgrowth of the Populist movement of the late nineteenth century and had been a key participant in that period's conflicts between farmers and financial interests. Thus, the Progressives in the Country Life movement and other supporters of extension viewed the Grange with suspicion and distrusted it as a vehicle for the new Cooperative Extension system. Despite the protests of the Grange and other established farm groups, then, Cooperative Extension was created with the express intention of beginning a new farm organization, one that did not come with preset political ambitions and class interests.[11]

Ironically, the farm bureau system did quickly become a forum for growers' political lobbying efforts, although in a different way than other farm organizations such as the Grange. Because the advisors themselves often came from the rural middle class and were required to have a minimal level of postsecondary education in some aspect of agricultural science, Cooperative Extension initially attracted more financially successful, educated, and relatively conservative farmers than did organizations like the Grange.[12] And though the architects of Cooperative Extension intended the farm bureaus mainly as centers to improve farm productivity and the standard of country life, growers quickly transformed the farm bureaus into a political and economic organization for influencing farm legislation on all levels of government. Once the county farm bureaus were established, farmers wasted no time connecting them into larger state and national organizations. For instance, California farm bureau members throughout the state formed the California Farm Bureau Federation in October, 1918 (Scheuring 1988, 26). The national farm bureau organization, the American Farm Bureau Federation (AFBF), followed in 1919.[13]

Once a farm advisor had organized a local farm bureau in his county, he was supposed to work with all farmers, whether they were members of the farm bureau organization or not. Most advisors surely did work with a broad segment of their communities, but they continued to have very close ties to the farm bureau structure, especially at the county level. In many states, advisors acted as administrative members of the county farm bureau, and conversely, some farm bureaus had a quasi-governmental status at this level. These close links between the two organizations brought cries of protest from the Grange and other competing farm groups, and prompted the AFBF and the national Cooperative Extension division within the USDA to sign a memorandum of understanding, officially marking off the boundaries and responsibilities between them, in 1921 (Scheuring 1988, 26). In practice, though, these close ties continued on an informal basis.

With funding coming from all levels of government and a very large grey area surrounding the extent to which farm advisors were part-time employees of the farm bureaus, it is not surprising that Cooperative Extension, like the larger land-grant system, developed an identity crisis with respect to its mission (McConnell 1953, 45). In some respects, the mission of Cooperative Extension was actually clearer than for university-based researchers: advisors were expected to work within just one county to improve farming practices there. The compatibility of this work with the larger aims of the land-grant mission, however, was more fuzzy. For instance, would increased productivity and efficiency lead to healthier farm communities? And, if so, what kind of farm was best for the health of these communities? Traditionally, the family farm was held to be the foundation for a healthy rural life and for U.S. democracy, but it was not clear whether knowledge and progress—at least the Progressive version of them—were appropriate goals for all kinds of family farms. These questions were especially important in the context of California agriculture during the period when Cooperative Extension advisors began to appear in the state's counties.

Cooperative Extension Comes to California

Extension came to California in 1913, with the same wave of anticipatory preparation that swept through many land-grant institutions just prior to the Smith-Lever legislation of 1914. The UC hired B. H. Crocheron, a student of Liberty Hyde Bailey, to head its new extension division, which at that time was called the Division of Agricultural Extension. Crocheron administered UC Cooperative Extension for 35 years before his death in 1948, and he put a stamp on the UC's version of farm advising that still exists today. For example, Crocheron insisted that advisors' salaries be funded through the university system, not through their local county governments. In this way, advisors were UC employees and were somewhat insulated from local political and economic pressures. Advisors' expenses for their office, travel, supplies, and other needs were to be provided at the local level, through the county government and contributions from growers. Thus, when the UC first began assigning advisors to specific counties, each county was required to make a financial commitment of $2,000 to cover the advisor's travel and office expenses. In addition, counties were required to organize at least 20 percent of their growers into a farm bureau association by the time of the advisor's arrival, with each grower contributing $1 per year to the cost of maintaining a farm bureau center for meetings. Before the Smith-Lever Act was even passed, several counties expressed interest in having their own farm advisor, and in 1913 advisors were placed in four California counties. By the end of 1917 the UC had placed farm advisors in 24 counties throughout the state, and the number of counties with advisors continued to climb, helped in large part by an emergency appropriation at the federal level to increase Cooperative Extension's activities in rural areas during World War I (Crocheron 1914; Scheuring 1988, 12–19; 1995, 80–82).

Despite the great amount of initial enthusiasm for extension work in California, advising often proved to be an uphill struggle. This was especially true after the war, when problems of land distribution, overproduction, and low commodity prices plagued the state's farm industries. These problems were prevalent throughout the nation, but they were especially pronounced in California because of the unique challenges of the niche market industries (see chapter 1). Many of the early struggles of niche industry growers, and farm advisors' attempts to address these problems, can be traced back to changes in the land tenure structure of California agriculture that began in the late nineteenth century. Prior to about 1890, California agriculture was typified by a relatively small number of enormous ranches. These unusually large landholdings dated back to the system of Spanish and Mexican land grants that defined land ownership before the United States took control of California. This system of land distribution allotted huge tracts of land to wealthy and influential petitioners.[14] When U.S. control of California began in 1846, *Californios*, the Mexican landowners who owned and made claims on some of the largest of the tracts, controlled many of the largest ranches. These landowners found it difficult to maintain their land in the face of increased white migration and the collapse of the Mexican cattle industry in the early 1860s. A particularly bad drought in 1863–1864 ruined many of the *Californios*, who had already been weakened by rising land prices (and the accompanying increase in taxes), the high cost of litigation, and predatory white merchants and financiers.[15] In many cases, although land was not transferred directly from *Californios* to white speculators, the Mexican land grants were divided among a few particularly crafty or influential white settlers.

Many of these large plots were initially used for "bonanza" wheat farming, wheat grown on a scale measuring thousands of contiguous acres. Wheat farming and cattle ranching were the main uses of these large holdings throughout the 1860s and 1870s, but at the end of this period the bonanza wheat growers found that wheat was no longer particularly lucrative and that continuous monocropping on such a large scale was rapidly depleting the fertility of soils. Many began to switch to more intensive orchard crops, especially oranges, peaches, almonds, and grapes. This switch seemed like a natural choice, given California's climate, but many of the largest landholders soon realized that their huge acreages were not an appropriate scale for fruit and nut crops. In addition, as some of these growers became more successful in their chosen niche market industry, their property values rose, increasing taxes. In response to these factors, the land and water companies established by these land barons and by the Southern Pacific Railroad (which also owned huge acreages of California land) began investing in irrigation and other infrastructural improvements to their landholdings, hoping to subdivide the larger blocks and sell individual plots in "agricultural colonies," many just two to ten acres in size (Liebman 1983, chs. 2, 3; Stoll 1998, ch. 2).

As a result of this subdivision, the period 1890–1920 saw distinct changes in the landholding patterns of California agriculture. At least on average, and only temporarily, the state's farms became slightly smaller in scale. The fate awaiting most of these new colonies of growers was often grim, given the low prices and oversupply problems that often plagued the first several decades of the orchard industry in California. If the colonists had money up-front to purchase a good piece of land with trees that were already bearing fruit, then they had a much better chance of surviving these problems. But other, less fortunate growers often had to wait a year or two before their orchards yielded a crop. If commodity prices were low when the crop finally came in, these growers faced financial ruin.

This was the agricultural ecology that faced Cooperative Extension during its early years in California, and advisors' expertise was not always well suited to the problems facing growers in the niche markets. Advisors often had little experience with these crops, simply because the growers themselves were creating the techniques needed to grow and market them on a large scale. Further, the advisors often could not help smaller growers who fared poorly under postwar market conditions of low prices and too many orchard crops. Advisors were much better prepared to address underproduction than overproduction.

How was Cooperative Extension supposed to transform agriculture in California? The tensions and uncertainties are clear in *Progress in Agricultural Extension*, a series of reports by the UC director of Cooperative Extension, B. H. Crocheron, who felt that Cooperative Extension was sometimes misunderstood and underappreciated in this period. Throughout reports that Crocheron wrote just before and after the onset of the Great Depression, especially 1927 through 1931, he appears torn between the needs and constraints of smaller growers and the demands of a market economy.

The subdivision of large landholdings gave small growers new opportunities, but these offered very mixed prospects for success. This situation provided farm advisors with an opportunity of their own: to address the problems of small farm communities and fulfill the mandates of their mission. Crocheron's feelings on this issue appear conflicted. Although he expressed concern about the plight of the colony farmers, he also chastised them for naïveté and the "scrambled mess" that California agriculture had become:

If only you could unscramble eggs it would make things easy. If you could only take the farmers off the bad land and put them onto good land it would help a lot. If you could take the fellows on the little acreages, consolidate them into economic farm units and let the surplus farmers work elsewhere; if you could grub out all the bad orchards and let the good ones fill the market:—all these would bring relief to agriculture. But you can't unscramble eggs. The land is divided and settled and planted. Nobody knows how to unscramble the mess. Apparently the only way is to let nature take its course. Economics adjusts itself in the long run. The bad land goes back into pasture; the little farms consolidate themselves through failure and despair; the bad orchards pass out in time.

But it's a hard doctrine. To anyone with a touch of human kindness—most of all, to the members of the Extension service who carry the welfare of farm people on their hearts—it's a heart-breaking business. (August 1927, 2)

This excerpt packs a lot of metaphors about farming and economic forces, and reveals the ambiguities that challenged advisors in California. Crocheron's use of the "scrambled eggs" metaphor and his desire to unscramble them point to the kind of progressive repair and redemption that inspired extension work in the first place. All the sentences that begin "If you could . . ." represent a hope for intervention, for a more rational California agriculture where the eggs remain uncompromised. But Crocheron can see no way out of this mess, despite the advantages of the latest agricultural knowledge. Further, his comment about "nature taking its course" naturalizes this state of affairs and portrays these small landholders as evolutionarily challenged in the ongoing march of economic progress. In all, this excerpt clearly reflects the major themes of scientific progressivism, where expertise could solve both technical and moral failings, but Crocheron also implies that the colony growers' failure was an inevitable outcome of economic processes.

Other excerpts from Crocheron's reports further confirm his ambivalence toward small niche market growers. On one hand, he clearly believed that smaller, less efficient farms were the cause of low commodity prices in California. In an essay titled "Common Sense," he rails against the lack of common sense in California agriculture: "It isn't common sense to believe that an acre of mediocre land will support a family of modern American standards" (November 1927, 3). On the other hand, he also expresses heartache for these growers' plight, emphasizing the connection between family farming and the land-grant mission:

Some people believe that the family farm is doomed; that agriculture is destined to become a business of big corporate units where people are all hired men on the farms of big corporations. We hope that day will never come. The Agricultural Extension Service was founded, not so the world may have more food, but so that the farmers of America may live and prosper. If corporate farming becomes the prevalent type in America it will mean that we have failed in our task. (October 1930, 1)

Crocheron's proposed solution was for growers to take advantage of the latest and best knowledge available to them, thereby raising productivity and profitability through efficiency. This push for efficiency was often accompanied by moralistic admonitions, where Crocheron emphasized that growers would have to show their own initiative to follow this advice. The following excerpts are just two samples of many such warnings:

To decrease costs, through better methods, is required only knowledge and ability. It lies within the individual power of good farmers. You don't have to wait for somebody else to do it, nor stall for collective action. It's a hard old doctrine; not everyone can apply it. But for those who have the opportunity and ability, better farming methods present the quickest and surest remedy [for decreasing costs]. (April 1928, 3)

If it is to succeed, the family farm must respond to the needs of the times; it must have some "get up and git" about it. The little farms must apply intelligence to their business, must promote the use and sale of their products, must be willing to spend money to make money just as does the big business man. It pays us nothing to weep and wail about hard times. Something must be done; and it must be done by the farmers. No one else will or can do it for them. (October 1930, 2)

Further, Crocheron argued, growers *could* make a living in times of low commodity prices—if only they would let their local farm advisor instruct them:

Not everyone is suffering. Some farmers . . . are making money. They are the men on the best land who are getting high yields by means of their land and good methods. Furthermore, they have enough land so that if the profit per acre is smaller than formerly, there is still enough margin to support the family. These fortunate farmers are more numerous than most people realize. They usually keep quiet so [they] are not easily manifest. The records in our office show, however, that they are not as infrequent as some people suppose. Many of them are willing to give large credit to the farm advisors whose advice they have taken and whose work has resulted in the maintenance of many a farm family which would otherwise have failed. (September 1930, 2–3)

Of course, he was right; Cooperative Extension could only do so much to help farmers become more efficient and make the best of their farms. But a crucial point in this advice is Crocheron's link between good farming practices, inspired by the knowledge of Cooperative Extension advisors, and the implementation of these techniques on good-quality, large pieces of land. By emphasizing the quality and size of the grower's farm, Crocheron implied that only the larger growers, with more capital for land and other resources, would be successful over the long term.

Creating a Context for Industrial Agriculture

For many growers, advisors' new, efficient methods were not transparently better than standard practices. Spending extra money and time to increase the efficiency and productivity of a given acre of land did not always seem like the best of ideas, especially in times of big surpluses and low commodity prices (Danbom 1979, 89–91). This was especially true during the early 1930s, when the Great Depression's effect on commodity prices took a toll on even the largest growers, and many sought new crops to plant or other ways to survive this long downturn in prices. Increased efficiency and productivity were a harder sell for Cooperative Extension during this time, and yet Crocheron stuck to this solution as the only practical way to improve the economic prospects of California growers. In this excerpt from a report in 1931, he emphasizes the importance of efficiency regardless of the size of a grower's farm:

For the last five years we have had people rise up to remark that this business of getting more pounds of butterfat per cow, more eggs per hen, more peaches per acre, is only making things go from bad to worse; that increased efficiency has created the surplus, and that the Agricultural Extension Service, by bringing increased efficiency has brought a surplus. . . .

Does increased efficiency necessarily mean the creation of a surplus? Is the efficient man the real surplus producer? We think not. We believe the "marginal man" produces the surplus; that if there were only efficient producers they would manifest their brains and ability, not only [by] economical production but also by studying probable price trends and adjusting their production to prospective demand. It's the ignorant inefficient producer who goes "hog wild" and plants over all-creation, thus creating the surplus. Brains, ability, information:—these three never injured any industry. The trouble with farming is that it hasn't enough of them. As one man recently said, "My ignorant neighbor is a menace to me." (April 1931, 3–5)

This quotation makes the moralistic character of Crocheron's Progressivism quite clear. Growers who did not have the brains and ability to increase the efficiency of their land were not only doomed, they were actually responsible for the poor market conditions. This argument also further naturalized the failure of certain growers; those who could not plan and adjust properly for hard times were ignorant and would be the losers in a struggle for survival.

Crocheron was likely correct about the relationship between efficiency and overproduction: if growers could get higher yields from fewer acres, they could keep fewer acres in production and still make money. The reality of this situation for many of the smaller growers, though, made it difficult to take advantage of this advice. In fact, after several decades of decreasing farm sizes, the period after the Great Depression saw a reconsolidation of agricultural land in the state, as many smaller growers got out of farming (Liebman 1983, ch. 3).

Farm advising during its first two decades coincided with the rise of industrial agriculture in the form that we know it today. Looking back, it is tempting to draw a direct line of cause and effect, to see Cooperative Extension as a means for Progressive elites and urban industrialists to project their vision and interests into the countryside. Farm advisors' work promoting efficient new production practices helped to create the technological and economic infrastructure for industrial agriculture. Their discourses of efficiency and expertise served to legitimate the consolidation of niche industry farms. While these discourses sounded like calls for transformation, Danbom's thesis about the moral conservatism behind Progressive ideals suggests that maintaining and retrenching social boundaries was also an important part of the movement. Cooperative Extension's paradigm of knowledge, efficiency, and progress suited small growers in some crops quite well, but it did little to alleviate the structural and economic troubles facing small growers of niche market crops.

At the same time, although Cooperative Extension brought a vision of modernity to agriculture, the roots of industrial agriculture in California were already in place. Cooperative Extension was itself shaped just as much as the farm industry during these early years, and Crocheron's reports show how advisors struggled to reconcile their mission of improving farming and rural life with changes to the structure of agriculture that seemed out of their control. The ambiguity surrounding farm advisors' mission as agents of progress and repair only exacerbated this tension, prompting advisors to search for a stable sense of their own work and its moral and political basis. Chapter 3 describes this process for the case of Cooperative Extension and the farm industry in the Salinas Valley.

3 A Niche of Their Own: Cooperative Extension's Move toward Specialization

The Salinas Valley's Progressive Growers

In chapter 2, I described the impact of Progressive thought on the creation and early work of Cooperative Extension, especially the value placed on expertise and efficiency in refashioning U.S. agriculture. Most histories of Cooperative Extension tend to emphasize the opposition that farm advisors faced when trying to implement this vision of a new and more rational agriculture in the countryside. In fact, the very titles of historical works on Cooperative Extension, such as *The Reluctant Farmer* and *The Resisted Revolution*, suggest tension and conflict as the predominant features of the grower-advisor relationship (R. V. Scott 1970; Danbom 1979). But during my conversations with advisors and growers in the Salinas Valley, I was often struck by their use of the term "progressive" to describe the fresh produce and other intensive niche industries of the valley. Advisors frequently praised local growers for their interest in research results and for the speed with which they adopted new and superior farm practices:

Soil/Water[1]: I think there's a lot of very progressive [growers] here willing to change if they see a benefit. I think you don't have to push change as hard here as you might in other areas.

Retired Advisor: We have progressive growers here. The managers of these companies are college graduates, they're sharp.

Other descriptions such as "advanced" and "mature" also often accompanied these characterizations, and I began to wonder at the meanings behind these terms. What did a progressive farm industry mean in the context of a history where growers were often portrayed as resistant to the science-based intervention of farm advisors? Salinas Valley growers seemed to defy the standard stereotypes. They were often cast as intent on making the best use of new knowledge and techniques for farming, and it was often implied that the advisors themselves needed to be forward-thinking and technically oriented just to keep up with the pace of agriculture in the valley.

The meaning of growers' progressivism is an important theme in this chapter. When advisors and growers describe the progressive character of the Salinas Valley farm industry, they are typically pointing to the power structure of the industry and the ecology of power that shapes the relationships between advisors and growers. Therefore, whereas in the previous chapter I focused on the discursive frames used by experts and Country Life reformers to argue for the repair of U.S. agriculture, here I treat the structural elements of the relationship between Cooperative Extension and the farm industry in more detail. How do farm advisors and growers work with each other, and what forces shape this interaction?

The answers are rooted in a long-term process of accommodation on the part of Cooperative Extension, where the work of advisors in the valley was literally remade in order to suit the research needs of the local farm industry. These changes involved the movement toward disciplinary specializations on the part of farm advisors stationed in Monterey County. This specialization has made advisors powerful; they are highly valued by the farm industry and are the beneficiaries of considerable financial and political support from the industry. Contemporary farm advisors in the Salinas Valley do not have an identity crisis; they have solved that problem through specialization and in the process created a kind of niche market of their own. At the same time, however, questions remain about advisors' focus on the production problems of the wealthiest and most influential growers.

In this chapter I present a historical account of how Cooperative Extension began in Monterey County and the Salinas Valley as well as how farm advisors moved toward a disciplinary-based system of advising starting in the late 1950s. Most of the data come from my interviews with growers and advisors, and highlight the political and financial relationships between Cooperative Extension and the farm industry. Although I focus on the work of advisors in one county—Monterey County—the process of accommodation to local farming interests is a common theme of the history of extension work in California and the United States. Ultimately, Cooperative Extension addressed its uncertain identity and mission through flexibility and accommodation. Farm advising was intended to be local work, and the institution has adapted to the particular conditions and demands of specific places and farm industry practices. These factors shaped the work of advisors just as much as their own work transformed the ecology for farming. The story of Cooperative Extension in Monterey County serves as an example of how state experts become enmeshed in local networks of power.

Cooperative Extension Comes to Monterey County

The first farm advisor came to Monterey County in the spring of 1918. UC Extension officials worked with the Salinas Chamber of Commerce to prepare for the advisor's arrival, securing the use of office facilities in the Salinas City Hall as well as a Ford car for the advisor's travels.[2] Despite the fact that the county's farm bureau was still not fully organized, Thomas C. Mayhew arrived in Salinas on April 15 and began work. An article from a local Salinas newspaper, titled "Farm Bureau of County Gets Splendid Start," describes the formation of the county's farm bureau after Mayhew had arrived, and it strikes an optimistic, progressive tone for future Cooperative Extension work in the county:

> T. C. Mayhew, the new county farm advisor, was introduced and made a favorable impression. The prompt organization of the county [farm] bureau the day after the arrival of the county farm advisor speaks well for the future of the farm bureau. It bespeaks efficiency, energy and success. The farmers are showing a keen interest in the progressive step and they and the farm advisor will co-operate heartily that the most benefits may be obtained from the establishment of an advisor and a county farm bureau; that not only the farmers themselves may benefit but the nation also at a time when increased food production is urgent and vital and therefore the problems of the farmer are particularly important.[3]

The writer is clearly tapping Progressive Era themes in this statement, with the discourses of efficiency and the need for increased production during World War I.

After the initial meeting to create the Monterey County Farm Bureau and elect its leadership, Mayhew began a busy schedule of meetings at farm bureau centers, assemblies of growers in different areas of the county, each with its own meeting place and regional director. One of these farm bureau

Figure 3.1
Farm bureau center in Lockwood, at the southern end of Monterey County,
California, ca. 1918. Local farm bureau members would meet here to discuss agri-
cultural issues. In front of the center is a farm advisor's car. Courtesy of Monterey
County Department of Parks.

centers, located in Lockwood, is pictured in figure 3.1. The work of a farm
advisor during this time required almost constant travel; in fact, a year-end
report from Mayhew to Crocheron recounted 77 farm bureau meetings,
348 farm visits, and more than 13,000 miles added to the odometer of
Mayhew's Ford. By the end of 1918, Mayhew also reported 600 members
for the county's farm bureau (Crocheron 1918, 39).

The early reports to Crocheron from Monterey County show a quickly
growing program of advising by Mayhew. His early work focused on
improved wheat and barley varieties, better management of livestock (espe-

cially important during the war years), and advice to growers in the county's nascent lettuce industry. Early in 1920 a committee of the Monterey County Farm Bureau's board of directors met with their county's Board of Supervisors to secure a $1,000 appropriation to hire an assistant farm advisor to work with Mayhew.[4] After a few assistants shuffled through the county in the early 1920s, A. A. Tavernetti, a native of Monterey County, was hired; he was promoted to farm advisor when Mayhew left for a statewide leadership position with UC Cooperative Extension in 1924. Tavernetti would remain Monterey County's head farm advisor for the next 34 years, until his retirement in 1958. In addition, Monterey County's first home demonstration agent, Mabel Eager, began work in August 1922. Eager began a child nutrition plan intended to raise the weights of farm children to "normal" levels. She also gave "Americanization" courses to Japanese men and women, intended to teach them "American customs of serving food."[5]

Overall, Cooperative Extension's first two decades in Monterey County mirrored the experience of farm advisors in other California counties: success with field crops such as wheat and barley as well as with cattle ranching and dairying, but less impact on the niche market industries that began to define California agriculture in the 1910s and 1920s. The niche for the Salinas Valley was fresh produce, especially cool-season vegetable crops such as lettuce, celery, broccoli, and cauliflower. Unlike the climate in California's Central Valley, where temperatures regularly exceed 100°F in the summer months, the climate in the Salinas Valley stays cooler and moister, created by ocean air currents drawn down the valley from Monterey Bay. Before 1920 the availability of these delicate crops had been seasonal and costly, but growers engineered a new system of year-round vegetable production in the Salinas, Imperial, and Santa Maria valleys, shipping their produce to points as far as New York via rail on boxcars packed with ice. The post–World War I era brought a huge increase in the demand for and production of lettuce and other vegetables. In 1916 a two-horse team made the first shipment of lettuce from Salinas to San Francisco; only fourteen years later, in 1930, Salinas Valley growers were shipping thousands of boxcars of lettuce to points across the country.[6]

The rise of the vegetable industry was full of technical challenges. A fresh head of lettuce is vulnerable to a wide variety of pests and is quickly perishable once harvested and transported to market. Because the industry was so new, however, advisors often found it difficult to provide expertise to vegetable growers and tended to focus instead on problems associated with grain and livestock production. Although research was conducted on issues related to vegetable production, Cooperative Extension specialists and faculty from the UC campuses were more likely to take the lead on those problems. Locally, the advisors were given commodity assignments that matched the distribution of the county's crop acreage but not the dollar value of the crop produced on this acreage. For instance, as late as 1954, the structure of agricultural extension work in Monterey County included a county extension director (Tavernetti), a livestock advisor, a field crops advisor, and a fruit and vegetable crops advisor.[7] This division of labor reflected the acreage generally used for these crops and livestock in the county at the time: approximately 130,000 acres of grain and field crop production, 100,000 acres of vegetable and fruit production, and over 1 million acres of pasture land for cattle and sheep.[8] The dollar value produced by these crops and livestock was distributed quite differently: about $10 million each for field crops and livestock and $73 million for vegetables and fruit (Tavernetti 1954, 2).

As a consequence, one farm advisor was assigned to work with all the vegetable and fruit growers, yet the financial and organizational resources available to these growers dwarfed those of growers of all other commodities. This advisor struggled to accommodate the research and educational needs of this industry, especially given that many vegetable and fruit growers were themselves quite knowledgeable about the special techniques and difficulties associated with large-scale production of niche market industry crops. In an interview with this advisor, now retired, he described to me the troubles he faced after his appointment to Monterey County in 1950:

Retired Advisor: When I came here . . . I was pretty much responsible for all the work done on the vegetable and the small fruits and [laughs] even the poultry. We had a 4-H man. We had a field crops man. And we had a livestock man. And that was pretty much it. . . . That didn't work out. And having such an extremely broad responsibility here—[with me] the only one really doing any significant work or having any significant responsibilities, in such a broad-based industry . . . as the veg[etable] crops industry and the small fruit industry—all I can tell is I was pretty much a total failure for quite awhile.

The frustrations this advisor faced when trying to serve the niche market industries of Monterey County were shared by the industry's growers as well. As County Director A. A. Tavernetti reached retirement age in the late 1950s, members from the vegetable industry began pressuring the UC to restructure and expand the Cooperative Extension staff in Monterey County to better accommodate the research needs of agriculture in the Salinas Valley, especially the vegetable industry. William Huffman succeeded Tavernetti as the extension director for Monterey County in 1958 and was charged with overseeing changes to the organization and style of farm advising in the county. These changes abandoned the more common method of commodity-based assignments—advisors' being assigned to a specific commodity or group of commodities—in favor of a discipline-based format. In this new mode, advisors were assigned to work in disciplines of applied agricultural science, including entomology (study of insects), plant pathology (study of plant diseases), weed science, soils and irrigation, and agricultural engineering. Two advisors working at the time of this transition were sent to UC Davis for graduate training in plant pathology and weed science, and three others were assigned to the county in the remaining disciplinary specializations.

As the weed science advisor who was trained during this change described the situation, the local agricultural industry felt that these specializations would be much more useful to them than the previous commodity-based advisor positions. The following excerpt highlights the influence of vegetable industry growers in this transition:

WeedsSci: And well the growers here, the grower-shipper industry if you will, were very progressive, in all candidness. This is maybe the part I should say not to tell, but I'm going to. We weren't used very much in those days—[the fruit and vegetable industry] would pick up the phone and call [UC] Davis direct. In those days, professors did a lot of applied work, and they had answers for these people. . . . This was mid-50s now. Probably late 50s. So, they were gonna retool, or reconvert this position to the specialization, and that's when they were thinking the entomologist, pathologist, and the weed science position . . . and the rationale was, these grower-shippers wanted that specific timely information. . . . So they [pause] put this squeeze play on the university.

CRH: So it was kind of from pressure from the—

WeedSci: So it was from grower pressure from the industry if you will. And so, we feel we had a great team that had built up.

Like the retired advisor quoted previously, the weed science advisor emphasized the inefficacy of the older commodity-based assignments for satisfying the needs of the niche market industry growers. He also noted the industry's motivations for this change: "specific timely information" tailored for the conditions of continuous statewide vegetable production. His characterization of the industry as progressive was meant to emphasize not only the interests these growers showed in research results but also the advanced state of the industry vis-à-vis the advisors' own knowledge and expertise. Given this progressivism, advisors themselves needed to be progressive as well, matching the industry in specialization and advanced techniques, in order to maintain Cooperative Extension's utility to growers. Other advisors also noted the intentions of the vegetable industry and their research needs, and suggested that advisors could never hope to give general advice to such a progressive industry:

Retired Advisor: We have progressive growers here. The managers of these companies are college graduates, they're sharp. They know a lot more about the marketing and the business aspects than any of us . . . farm advisors. [CRH laughs.] Where we really performed a function, was that we knew more about diseases and insects and soils than most of them did. But in a business sense, we didn't try to advise 'em too much [both laugh]. . . . I think that would be vastly different in other areas. But here, they're a specialized industry.

CRH: It's because it's so much of an industry then that makes it different?

Retired Advisor: Yes, it's very intense.

PlantPath: In terms of our office, as you know our office is fairly unique. I guess not absolutely unique but, fairly unique in that our staffing is based on disciplines not commodities. . . . I think it's a good match, because we have a very progressive, uh, advanced mature industry, so, in some ways they don't need a lettuce farm advisor here. Because, what would a lettuce farm advisor tell a lettuce grower—how to grow lettuce? I mean, we learn from them. And they're so much ahead of us that, [the] university cannot hope to offer that. . . . Our positions are more specialized—you know [I'm] just doing plant pathology and it's real narrow but specialized. [The growers] don't have their own plant pathologists. So, we have a niche to fill there.

In these excerpts, the advisors emphasize their own specialization, their participation in a kind of niche market of their own. Their characterization of growers as progressive—they do their own research and are in many aspects of vegetable production and distribution more knowledgeable than advisors and other UC researchers—is tied to advisors' own narrowing of the focus of their work. Perhaps the most telling statement comes from the plant pathology advisor: "What would a lettuce farm advisor tell a lettuce grower—how to grow lettuce?" With this question he distinguishes himself and other advisors working on problems associated with lettuce production from the standard meaning of what it means to be a farm advisor. Accordingly, farm advising in Monterey County is portrayed as a mode of work to which the old rules should not apply. By describing growers as progressive, the advisors justify the way that their extension work has been ordered in Monterey County.

Specialization and Advisors' Ties to the Farm Industry

In general, advisors who serve under the discipline-based model believe that this is the best structure for extension work to accommodate the research needs of the commercial growers in their county, growers working with niche market industry crops on a relatively large scale. The switch to a discipline-based system of extension work in Monterey County helped to solve some of the identity problems that plagued Cooperative Extension from its inception, particularly in California. By focusing on, say, the diseases of fruit and vegetable crops instead of any and all problems associated with the production of lettuce or a subset of such crops, the advisors can tailor their program of work to specific problems facing the local farm industry. Three of the main advisor specializations that were implemented in the change to discipline-based advising involved pest control: insects, diseases, and weeds. The timing of this shift in the late 1950s corresponded to the availability of new and powerful synthetic pesticides for each of these types of pests, and growers in the Salinas Valley remain intensive users of these technologies. In this way, advisors' work was reshaped to conform to the practical and technological structure of the vegetable industry, so that growers could receive timely information that helps to maintain a complex system of farm production.

This new system provided advisors a way to work out their mission, but not without consequences; the move toward discipline-based advising has changed the kind of work that advisors do, and who they do it with. With their specialization and emphasis on research, Monterey County's farm advisors are less able to provide the traditional basic extension advice needed by the more modest growers whose plight inspired the creation of Cooperative Extension. Although Monterey County no longer has any substantial dairy or field crop growers, there are still growers working in strawberries and other small fruits, organic vegetable production, and other areas of farming where growers can make a humble living on a much smaller scale of production than the county's larger commercial growers. These two groups represent very different sets of clientele for advisors in the Salinas Valley, and their choices about which group to focus on usually have trade-offs associated with the capitalization of their clientele and the kinds of help they need. Advisors sometimes find it difficult to reconcile their program of extension work with the varying needs of large and small growers. The county extension director described this dilemma in terms of the current advisors' job demands:

Director: The entry-level farmers need somebody to be out there, watching what they do every week, the old-time farm advisor way. And advisors now simply don't have the time to do that.

Smaller growers, especially beginners with basic production questions, often require very basic advice on, as another advisor told me, such questions as, "What's this?" "What's that?" "How come all my plants are dead?" This is not to say that they cannot benefit from research on their crops and farming practices, but they may not be in a position to use this information until they master actually growing a successful crop. Ironically, much of this work with smaller growers is truly advising—giving less experienced growers advice on basic methods of crop production. But as a consequence of the changed emphasis, the meaning of being an advisor has changed, and the main aspect of an advisor's work has become research and problem solving.

Those small growers who *do* have considerable experience with crop production may yet have very serious farming problems—most often rooted in the economics of agriculture—that are difficult to solve. In fact, not only are these small growers' problems different, they are usually much more complex and less likely to be solved through a quick fix, such as the addition of a new input or a modest change in practices. Much like the early days of Cooperative Extension in California, when Crocheron struggled to justify advisors' work toward increased production and efficiency, these problems of marketing and distribution are typically beyond the scope of advisors' expertise. The weed science advisor made some interesting contrasts between his discipline-based work in weed science and the other component of his position, which was to aid the county's field crop growers. His observations are especially telling because his own program represented the two domains of extension work discussed here: research and advising. This sequence began with the advisor trying to think of particularly hard problems, and he contrasted his work in weed control with his attempts to improve the economic situation of field crop growers during the 1960s:

WeedSci: I used to always joke with my friends, off to the side, because we didn't want to let our bosses know but—weed science was actually a good discipline because it was so dramatic. You either controlled those weeds or you didn't. And so when you developed something, it really was such a hero-looking type of concept. . . .

Um . . . I'm trying to think of other, getting back to the question [of hard problems]. One of our goals was to improve the economic structure of the dry-land barley growers. In South County we had about 100,000 acres that was dedicated to what we call summer fallow, where they grow one crop every two years. . . . So you had two years of moisture theoretically to grow the crop. Well, the people who grew it also had cattle operations [and] barley was their cash crop, outside of the livestock. At the best you got 20 sacks [per acre]—in those days you got $5 a sack, $100 an acre gross. It was pretty marginal. Of course a lot of them owned their land so the rents were lower. And the inputs were relatively low—mostly soil tillage, tractor work, et cetera. And most of them were family farms and the son or daughter were on the tractor. But, my objective was to . . . develop another crop that would have more revenues, so part of my [objective] was always looking for new crops. And I'd go down there and look at things that might have some appeal. We looked at things like canola, safflower—some of these oilseed crops. We even tried soybeans. Anything off-the-wall we'd go down there and try. And to be honest with you, none of them succeeded.

CRH: It's kind of funny: it sounds like, even though the weed science was kind of a new area when you started on it, that in a lot of cases the weed problems were not quite as complicated as some of the ones with the agronomic [field] crops, do you think?

WeedSci: That's true. I think it was a greater challenge in the agronomic arena. Part of it too was you were working on narrow margins of profit in the operation. So you didn't have the flexibility of introducing a $50 an acre herbicide. . . . In this grain crop, you didn't have that latitude to work with.

This seems to contradict the statement by the retired advisor in which he noted the progressive character of Monterey County growers and said, "In a business sense, we didn't try to advise 'em too much." Ostensibly, the advisors needed to specialize to match the progressivism of the growers and the especially intense research needs of their industry. In many ways, however, the needs of the small growers described here by the weed science advisor were much more complex and potentially intense than those of the larger growers.

Each of these factors contribute toward contemporary farm advisors having less and less time for actual "advising," and orienting their extension programs more and more toward problem solving and research. Perhaps the biggest factor that influences the direction of their work stems from their financial relationships with the local farm industry. The disciplinary mode of advising suits commercial growers quite well, and the Monterey County farm advisors have prospered under these growers' appreciation and largesse. For example, in the early 1990s, the advisors moved into a new office and research facility built at county and farm industry expense, and commercial growers were very influential in decisions to include laboratory and greenhouse facilities in the new building. With these facilities in place, the advisors now have access to diagnostic and research tools that advisors in most other California counties do not have. In most cases, UC farm advisors need to send plant or soil samples to UC Davis or another of the UC's agriculturally oriented campuses for analysis. With these new facilities in Monterey County, advisors there can now do much of this work in-house.

Similarly, advisors in Monterey County can tap into a relatively large amount of farm industry money earmarked for applied research grants. Under the terms of the California Marketing Act of 1937, growers are allowed to organize special industry boards around a given commodity, such as lettuce, tomatoes, raisins, or any other crop whose growers have the requisite capitalization and self-organization to form a commodity board. These commodity boards are closely regulated by the California Department of Food and Agriculture and typically focus their efforts on marketing, advertising, or research. For instance, the most powerful and influential commodity board in the Salinas Valley, the California Lettuce Research Board directs its attention solely toward research problems associated with the production of lettuce. Based in Salinas, the Lettuce Board assesses a tax on all the entities in California that produce lettuce. It collects a fraction of one cent per carton of lettuce, but these pennies add up: from the inception of the board in 1973, through 2001, the board invested about $9 million in lettuce research, focusing on issues related to lettuce breeding, pest control, postharvest handling and transportation, and other production-oriented problems of lettuce farming (CLRB 2007).

In practice, preference is given to applications from UC and USDA researchers, although some funding is given to scientists in other agencies or with private groups. The board tends to fund proposals from researchers who have a proven track record of successful research on problems associated with lettuce production. These preferences have made Monterey County advisors the beneficiaries of many board grants over the years because they are local researchers who can devote a considerable amount of their research activities toward applied production problems, especially in the area of pest control. Further, the advisors' proximity to the main players in the lettuce industry (and thus the Lettuce Board) makes them familiar to those making the funding decisions. The entomology advisor commented on his relationship with vegetable industry commodity boards:

Entomol: I have research projects with the Lettuce Board and the Celery Board. They've been funding me ever since I got here in '87 pretty much. . . . And they have a philosophy I think of making a commitment both to problems that they have and to certain researchers who they know will work on their problems. And so there are a group of us that are kind of their core people that they fund on a regular basis. And that does depend on the research projects and the proposals you submit—but, you know I'd have to screw up pretty badly for the Celery Board to say they weren't gonna give me any money next year. Because they've kind of made a commitment to me as one of the people that they're going to work with.

CRH: You have like a working relationship with them.

Entomol: Exactly. So, there are probably half a dozen of us that are in that category for celery and probably a dozen in the Lettuce Board.

Thus, there is an expectation that advisors will work within the structure of the commodity board system, and with respect to agricultural science this makes the Lettuce Board the proverbial 800-pound gorilla of the Salinas Valley. Paying attention to one or another problem means aligning oneself with the concerns of those who *have* the problem. These decisions always represent compromises, and one advisor explicitly framed these results in terms of the local politics of agricultural research:

CRH: Would you say that extension advisors have to worry about or deal with politics in any sense?

Advisor: Yes, we do—it depends on what you mean by politics. If you mean political parties, no. But if you mean the reality of politics within industries, yes, you do. I mean, you can't be totally blind to that. For example, if I were trying to conduct a program here and not work in lettuce, per se, and not work within the structure of the Lettuce Board, that would not be acceptable. It's a major crop, you know, it's where most of the power base is for the growers. . . . So politically it would not be acceptable for an advisor to not work with lettuce.

Larger growers also tend to be much more vocal about their research needs, and their connections to commodity boards and research funding allows them to get the attention of researchers. Although the commodity boards are the biggest and most formal source of funding for advisors, more informal arrangements are often made to deal with problems specific to a minor crop or a special circumstance. By pointing to these problems and offering money for trials, larger growers can also attract the attention of the advisors. The plant pathology advisor described one such situation:

PlantPath: If there's really something going on that you don't know about already . . . they'll let you know. They'll say, "Have you been following that new strain of mildew on the lettuce?" or, "What do you know about that?" If you're not aware of it you hear about it, and it gives you a chance to see if there's a need to get involved.

The thing with cauliflower verticillium . . . it was one of my first big projects when I came here in 1990. And the manager for [Company X] just called me and said, "I want to show you this field. We think it's something unusual and it's wiping out the whole field—it's just turning yellow. Will you take a ride with me?" And, so I went out for a ride and that turned out to be the start of a project that went four years and we got a pretty generous amount of funding—probably close to $40,000. Which is—that's pretty good for me.

CRH: Does cauliflower have a [commodity] board?

PlantPath: No, it was informal. It was a big enough problem to where the major cauliflower growers just funded it. Just contributed to a pot. So it resulted in a nice project. Got some publications out of that and identified a new problem for them. It came out of an informal encounter.

With advisors' move toward specialization and the amount of money made available to them for research projects, working on the problems of large growers is not a question—one simply makes sure to work on their problems in order to remain in the good graces of the commodity boards, but also, of course, to impact the growers who are farming the majority of the land in the valley.[9] Thus, the disciplinary structure of farm advising in Monterey County, the availability of research money, the nature of commercial growers' problems, the publication demands of the university, and the intractability of many problems facing smaller growers all contribute to advisors' tendency to spend more time on the research and problem-solving needs of large growers.

Critiques of Cooperative Extension's Ties to Industrial Agriculture

The move toward research-based extension work has been challenged, both by contemporary advocates of the agrarian, small-farm ideal and by critics who charge that the UC system, particularly Cooperative Extension, discriminates by ethnicity and class in its selection of clientele. The publication of Jim Hightower's *Hard Tomatoes, Hard Times* in 1973 launched a renewed debate on the mission of the land-grant schools and state-funded agricultural research. Hightower accused campus-based researchers and extension personnel in the nation's land-grant universities of focusing exclusively on the problems of "agribusiness," and he argued that this attention had led to a decline in small family farms in the United States. Similarly, other critics, some of them UC researchers themselves, questioned the UC's agricultural research priorities.[10] These writers examined the "social consequences" of agricultural research and depicted university scientists, especially those involved with the mechanization of agricultural production, as major contributors to the increasing concentration of farmland and production in the hands of a few large industrial growers. One farm group even brought a lawsuit against the UC, charging that its development of the mechanical tomato harvester in cooperation with equipment manufacturers and farm industry groups constituted a conflict of interest and a violation of the Hatch Experiment Station Act's mandate for land-grant research. Other groups brought charges of racial discrimination against the Cooperative Extension system, both with respect to its hiring practices and advisors' clientele. Lawmakers in the California State Legislature questioned the programmatic priorities of Cooperative Extension and the UC's record on hiring minorities and women for advisor positions (Scheuring 1995, ch. 7). These critiques led the UC to establish a number of new programs and policies intended to address these problems. In this section I describe some of these actions and the responses to them from advisors, commercial growers, and advocates for small farmers in Monterey County.

Following these criticisms in the 1970s, the UC developed affirmative action policies for Cooperative Extension hiring and encouraged advisors to develop new clientele through special outreach programs. The university created a formal accounting procedure to evaluate individual advisors' efforts in this area and requires them to provide tallies of the different racial and ethnic groups they serve in a given year in their county. This figure can then be compared with census figures or other data that provide an overall idea of the population that could potentially be served. This form of evaluation is a major component of each advisor's yearly reviews and the formal review process for their promotion within the UC system. In addition, new efforts at outreach to small growers were initiated through the Small Farms Center, an organization formed on the Davis campus in 1978 to generate and distribute information tailored to the needs of small growers. Several farm advisors with small-farms assignments were also placed in counties throughout the state. Finally, the UC and the state sponsored several programs for educating and providing financial assistance to small growers, especially in the strawberry industry. Former field-workers, many of them Mexican-Americans, took advantage of these programs and formed *cooperativas* to produce and market strawberries in the Santa Cruz and Watsonville areas. In all, there was flurry of activity in the UC during this period, as the system scrambled to respond to critiques at the national, state, and local levels (Scheuring 1995, ch. 7).

In my interviews with the now-retired cohort of Monterey County farm advisors—the advisors who were first impacted by these new programs—their reactions were mostly negative. By the late 1970s, when the UC began encouraging advisors to work more closely with minority populations and small growers, Monterey County advisors had been working in their disciplinary specializations for nearly 20 years, and their focus on research-based service to commercial growers and the commodity boards was well established. As white men, they seemed puzzled and even offended by calls to work proactively with new clientele groups:

Retired Advisor: A lot of us old-timers got frustrated in this era when we did start getting some inputs that we weren't spending enough time with minorities. And our natural rebuff was, "Hey, the door's open there. Anybody can walk in and we're not gonna refuse 'em." And that was always our retort, that we didn't feel we had to be proactive and go and knock on doors of a Japanese guy or a Finnish guy or an Armenian [laughing] or a Hispanic guy. We were here, and, we liked to think, a grower acknowledged that, hey, there's a source of information. Now the only—trying to look realistic—the only criticism might be, is that we had a language barrier. Yes, maybe they never heard of extension, and particularly if they came recently from the Third World or another country. Uh, but the traditionalists knew who we were. Even the new guy starting out, he might not be [on the] magnitude of [Big Grower X] or [Big Grower Y], but ... they knew who we were and they would come and ask for information.

Retired Advisor: There is still a general, good respect for university work here ... for a lot of the university work. You expect agriculture to be politically, socially quite conservative, and they're not in favor at all of a lot of the social programs that the university has gotten into. Diverted a lot of resources over to that. So [commercial growers] don't appreciate that.

Extension has made changes along those lines, and extension here [in Monterey County] has not changed as much in that regard as some other areas. We've remained almost exclusively agriculturally oriented, with the exception of the home advisor program. There's been some pressures—in the last few years [of my career] there was a lot of pressure for us to develop all sorts of programs in these areas, even on us [farm advisors], which I thought was kind of foolish. So we would write up a few things, but in reality, we stayed pretty much—we knew where our priorities *had* to lie, and where our support was coming from. . . . The new [advisor] coming along has got a little bit more difficult job, I think. Because his advancement is at stake—he's got to show this and show that. And a lot of those figures that went in all over the state, I think a lot of that, a lot of the social programs, low-income programs—not so much with the home advisors but with those of us in the ag field—there was just an awful lot of smoke with that stuff that went in. It was a lot of effort, a lot of planning, a lot of paperwork that, I think, for all intents and purposes, just stacked up in some room and never was looked at again. A lot of time and effort and money went into that, and I think, with very little good, practical results.

According to these advisors, the door was open to all, but they could not be expected to make radical changes to their programs in order to reach out to new clientele groups, particularly if this took them away from work with larger growers. They emphasized that they were willing to help smaller growers—the weed science advisor's discussion of his work with barley growers in the 1960s is just one example—but that growers had to show some initiative of their own. However, under this scenario, not all ethnic groups of small growers could take equal advantage of extension resources. In a study of Monterey County strawberry growers, Daniel Mountjoy (1996) found that Anglo- and Japanese-American growers were more likely to cite farm advisors as trusted sources of information on farming practices than were Mexican-American growers, who were more likely to cite respected growers in their peer group. Although language was sometimes a barrier, Mountjoy's analysis shows that cultural differences among the groups constituted the most important distinction. Given these differences, expecting growers to actively approach advisors only exacerbated inequalities in the extent to which underserved grower populations could benefit from extension services.

In addition, these advisors' assertions about an "open door" veils the extent to which they actually resisted the new programs. Many of the retired advisors simply viewed these programs as a waste of time and used their support from commercial growers to cover this resistance. In some cases, the advisors also voiced a kind of evolutionary argument to rationalize their actions. Just as Crocheron had lamented the "scrambled eggs" pervading California farming in the early days of UC Cooperative Extension and predicted that economics would eventually weed out disadvantaged growers, one retired advisor claimed, "Ever since the beginning of time, a trend has been taking place in agriculture . . . with a consolidation of a lot of farming interests into a few, large growers." In this way, the advisors could justify their opposition to the new programs and limit their efforts to half-hearted attempts at satisfying the new affirmative action requirements.

These attitudes likely represent the same kind of social conservativism that a retired advisor attributed to (white) industrial agriculture in the excerpt above. This advisor emphasized the support the industry had given to extension work and how Monterey County was lucky to remain in the good graces of the farm industry. His stress on this point reveals the danger advisors felt, and still feel, in jeopardizing their relationship with the farm industry by spending more time with small growers. In my interviews with commercial growers, I found that they held advisors in high esteem but were in fact suspicious of the motives of the UC system. Although the growers disapproved of the UC's "social programs," they attributed them to the political machinations of UC administrators, not farm advisors:

Grower: I have this problem with some of the politically correct crap that comes from Davis and Berkeley. . . . I think [Cooperative Extension] has been diverted—and I'm not a bigot at all—but I think that extension has been diverted into *cooperativa* farms in this county and in Santa Cruz. People are spending a hell of a lot of time out there with a guy that grows an acre and a half of strawberries. They're funded, but they don't make any money—they get government funds to grow it and extension people to show 'em how to grow it, and it's ineffective as hell. The minute you pull the support out from under 'em, you know, that's it. And I'm not bigoted against Hispanics—believe me. We have 99 percent Hispanics working for this company and 98 percent of 'em are damn good people. But I think that's some of the areas that [the UC gets] a little goofy on. And I think it comes from the top.

CRH: Does that ever provide a problem here on the local level in Monterey County, or is it mostly confined to the talk up there?

Grower: Well, it diverts time and resources. . . . Some of [the farm advisors'] time was being diverted into stuff that was not economically viable and was somebody's pet idea on what oughta be done. And, if I didn't think it was a cynical, politically correct, way to get-my-budget-increased kind of crap, it wouldn't bother me. You know if a guy says, "You know what? I can make 15 migrant Mexican families stay home, and raise strawberries and make $30,000 a year, and that's good for society"—I can accept that. If he genuinely believes it. But when it's, "Just do that crap so nobody gets on our butt. It's the right thing to do for the politics at the moment"— I don't like that kind of cynicism.

This grower emphasizes the economics of small farming and frames the UC's involvement in such undertakings as a cynical waste of resources. Several other growers echoed these feelings, stressing the larger economic importance of industrial agriculture. These commercial growers acknowledged that Cooperative Extension had a mission to serve the general public, but they also argued that the UC needed "to spend a bigger chunk of time with the problems that our industry has, because our industry is driving the economy."

Influenced in part by these feelings within the farm industry, the now-retired cohort of advisors in Monterey County chose to ignore pressures from the UC to expand their clientele base, and the strong support of commercial growers undoubtedly helped them maintain this resistance until their retirement. This cohort of advisors who oversaw the changeover to discipline-based advising in the late 1950s and 1960s retired in the 1980s. The new group of farm advisors hired to replace them began their careers as the UC was experimenting with and formally implementing affirmative action policies and were not in a position, with respect to their career trajectory, to simply resist these new programs. In addition, the current advisors were less likely to express the same kind of conservativism toward working with the groups targeted through these initiatives.[11] A new county extension director, appointed in the early 1990s, has been especially active in making sure advisors are connected to these groups and that small minority growers know how to use Cooperative Extension's services. For instance, during the time of my fieldwork with the advisors in Monterey County, the director was running a night course in bookkeeping to help small growers keep track of their farm operation's cash flow and also worked to produce a bilingual (English and Spanish) outreach video to educate growers about the services available from Cooperative Extension. Similarly, some of the farm advisors worked with members from local organizations that served as training and advocacy groups for small growers.

Despite these differences, advisors still work within the same disciplinary-based system and have very strong ties to commercial growers and their research funds. Just as the retired advisors worried about the perceptions commercial growers would have if they spent a lot of time working with new clientele groups, the current advisors remain sensitive to the politics of the industry and the kinds of ties that keep them connected to industry-based sources of funding. The advisors often find it difficult to reconcile these divergent demands, and the small-farms clientele usually gets less attention as a result. Recall the advisor's statement describing the politics of working on problems of the lettuce industry and how it would be unacceptable for him not to conduct research on lettuce.

Continuing Uncertainty: Cooperative Extension and Its Mission of Repair

More than 20 years after the criticisms leveled at the UC during the late 1970s, the clientele of Cooperative Extension remained a source of controversy. To find out more about the perspectives of small growers, I talked to some local advocates for smaller, minority farmers. These advocates administered groups that educated and organized those interested in beginning their own small farms, with memberships that consisted almost exclusively of Mexican-Americans, most of them former farmworkers. In my discussions with these advocates, they emphasized the educational needs of these small growers and talked of their efforts to enlist the support of Cooperative Extension advisors. Although their organizations had benefited from meetings, workshops, and field trials performed for them by advisors, they felt that commercial growers still received most of the advisors' attention, and called for change in this regard. This advocate, for example, expressed anger and frustration over failed attempts to get basic support for growers in the advocate's organization, questioning advisors' adherence to the land-grant mission:

CRH: So, I wonder if you have a general sense . . . of what it is that the university should be doing. . . . First of all, do they have a responsibility to the growers in your organization, and, if so, what should they be doing—what's their mission to the growers in [your organization]?

Advocate: First of all, to answer that question, I would have to ask my own question [of Cooperative Extension]: "What's your mission? What *is* your mission? After all these meetings [with Cooperative Extension], what is your mission?" We've worked with them to give them advice, we've given them ideas to develop certain topics. From that, what have they gathered? "What's your mission for [our organization]?"

I would like to revisit their goals and objectives, just to see what they're supposed to be doing. Who was their constituency supposed to be—what is the initial purpose of developing Co-op Extension? My understanding is . . . that they're supposed to help the disadvantaged grower, the ones who are in need of specific information. We should come back to that—we should go back to the goals and objectives of their initial development.

Another advocate voiced similar concerns but struck a more optimistic tone about advisors' attempts to make better connections with small growers:

CRH: I wanted to ask if you have any feelings about things the university could be doing to promote the kind of work that you'd like to see in agriculture in this area. What kinds of things do you think the university could be doing? Or are they doing everything that they could be?

Advocate: There is always more to do, but I do have strong feelings about certain things . . . that I feel that the university system can do. One is to balance the ship [holds up hands to represent two sides of an unbalanced scale] and say, there is this ship where a lot of money is here, and a little money here. Here, this is where we are [indicating the lower side], non-profit organizations, for example. Or, an organization that is helping people to become farmers or that is dealing with small farmers. We need to find ways and encourage the system to share resources.

I spoke to these advocates 80 years after Thomas Mayhew first came to work as a farm advisor in Monterey County. That they could still question the responsibilities of Cooperative Extension and ask advisors, "What *is* your mission?" is a remarkable example of the persisting conflicts over what constitutes the public good for agricultural science and the land-grant system. Cooperative Extension's mission crisis remains unresolved despite advisors' attempts to strike some kind of balance between the diverse needs and opportunities represented by different types of clientele.

My discussion in this chapter has focused on the ambiguities in Cooperative Extension's mission and the consequences of this uncertainty. Extension work was intended to mend the broken social structures and economic conditions of rural life, but no one seemed to agree on what this meant. Science and rational planning were to provide the answer, but what was the question? The conflicts surrounding these questions makes an ironic point about advisors' work: though Cooperative Extension was intended as an institution of change and repair, the meaning of their own mission has also been the subject of negotiation and intervention. Just as repair is a site for power relations to manifest themselves in other areas where orders are constructed, so Cooperative Extension itself has gone through this same process. Mission ambiguity has allowed a whole host of competing influences to shape the direction of extension work, and it is still being negotiated. Some of these influences were, in a sense, global, and built in to the design of Cooperative Extension, especially the farm bureau structure and its early effect on the clientele of advisors. Many other factors, however, have been much more local, especially the power of local actors to collapse the ambiguity surrounding extension's mission into an express set of means and ends.

In Monterey County this process has led to discipline-based advising, and it is clear that the power and influence of the niche market industries has shaped the kind of work that advisors may choose to do. By the same token, small growers have also shown their influence through advocates at the county and state levels. Advisors' attention to their problems, however, is limited when compared to the service they provide commercial growers, pointing to limits on the negotiated character of advisors' work. Although one could argue that the affirmative action and small-farms programs implemented by the UC in the 1970s were attempts to change the direction of advisors' work, these changes were clearly not transformative. The farm industry and the advisors themselves largely maintained their relationships in the Salinas Valley.

Finally, just as the commercial growers in Monterey County have had a strong influence on the direction of advisors' work in this locale, so this process has made Cooperative Extension look different in other places. In fact, in some cases, Cooperative Extension has taken precisely the opposite approach. Instead of specializing in the problems of commercial agriculture, other counties have focused on small farmers, home gardeners, and community development, often because there are simply no commercial growers to support the kind of research-based advising that advisors in Monterey County perform. This alternative path has been taken in other California counties and in counties throughout the nation. In a sense, this diversity is an empirical confirmation of my argument about the politics of repair; extension work both shapes and is shaped by the local ecology in which farm advisors find themselves living and working.

4   A Repair Crisis: Farm Labor during World War II and Beyond

An Ongoing Conflict

As a small agricultural community, Salinas does not seem a likely place to find protest and violence. Periodically, however, the town has been a battle zone, as long-simmering tensions led to public conflicts between growers and farmworkers. In the fall of 1936 a strike action by lettuce packers in Salinas captured headlines across the nation. Thirty-four years later, César Chávez's United Farm Workers shifted its focus from the grape industry of the Central Valley to the Salinas Valley's vegetable industry, sparking another major battle between growers and farm labor. In this chapter I use the strikes of the 1930s and 1970s to frame a period of conflict that drew growers and their backers, farmworkers and their sympathizers, and scientists and other state-based resources into a complex struggle to maintain or transform the social and material order of California agriculture.

In each case, these strikes violently and publicly revealed conflicts that had always been at the heart of California's farm industry. For California's niche market growers, farm labor has always been a very important part of their production practices, and growers have long struggled to control the use of farm labor. Even today, when much U.S. agriculture is mechanized, many niche market industries remain highly dependent on the use of seasonal, often migrant farmworkers to raise and harvest crops. This labor system allows growers to produce niche market industry crops on a scale and at a cost that can make these crops very profitable. Growers' dependence on labor, though, has proven to be a double-edged sword because they have constantly struggled to maintain control over labor in the face of unionization and other challenges to their farm labor system. Accordingly, conflict between growers and farm labor has been an ongoing feature of niche market farming in California, and the process by which actors have defined the roots of the problem and proposed solutions is an integral part of this history of conflict. Growers, farm labor, and the state each defined farm labor as problematic for California agriculture, but their definitions differed in fundamental ways, and these differences, in turn, suggested divergent forms of institutional repair.

Growers and other farm interests, for example, often defined the problem in terms of the reliability of the labor supply—having a large enough supply of workers to keep wages low—and ongoing attempts by workers to unionize. Growers chose multiple methods to address these perceived problems, including organizing to prevent the unionization of farm workers, soliciting support from all levels of the state, and controlling public opinion through the media. These solutions, however, constituted a maintenance approach to repair. Although growers' solutions involved a great deal of organizational work and cooperation with the state, they changed relatively little about how growers actually interacted with workers, and they served to consolidate growers' control over farm labor. Farm labor was a *production* problem for growers, and state experts, including UC agricultural scientists, often saw labor conflicts in these same terms.

In contrast, critics of California agriculture's dependence on and exploitation of farmworkers pointed to the continuous cycles of labor conflict and argued that the system was inherently irrational and required a more humane and labor-friendly form of farmwork. These critics, such as the writer Carey McWilliams and the UC economist Paul Taylor, defined labor conflicts as a *social* problem and called for broad change during the labor unrest of the 1930s. Their more systemic, or transformative, calls for repair included plans for improving the working conditions of farm workers and the suggestion (by McWilliams) to break up California's large farms into smaller ones.

These two approaches to repair—maintenance or transformation—determined actors' responses to the labor demands of World War II. The war was fought through the power of war commodity production, and labor was an integral part of this mobilization. Just as the government took special measures to increase factory production, it also acted to increase agricultural production during the war years, and many of these efforts were centered around the issue of farm labor. The onset of U.S. involvement in the war took workers away from manufacturing, and this was also true in agriculture—perhaps even more so. Therefore, the impact of the war appeared to plunge California agriculture into a farm labor crisis, but the problems growers faced in this period were actually rooted in the ongoing history of labor conflict within the state's farm industry. Although the war appeared to temporarily dispel conflict as the country pulled together for the war effort, the meaning of the farm labor crisis in California and opinions on how to deal with it remained contested, showing a great deal of continuity with prewar labor conflicts. These discourses, in turn, shaped the structure of California's farm labor system in the two decades following the end of World War II.

Through these different visions and experiences of farm labor before, during, and after the war, I show how discourses about the rationality or irrationality of repair strategies are deployed in times of social change, and how actors construct the meaning of a crisis itself to emphasize the need for certain forms and levels of repair. Throughout this chapter I emphasize the role of the state for understanding how the farm industry maintained a system of production despite threats to its stability. Growers received a great deal of help from the state in their struggle to control farm labor. A large literature on the history of farm labor in California has emphasized growers' access to political support (through grower-friendly legislation and court decisions), financial resources (through banks and other organizations with a financial interest in the farm industry's profitability), and violence (through police intimidation).[1] Much less has been written, however, about the state's intervention beyond legislation, the courts, and the police, such as the role that organizations like Cooperative Extension played in conflicts over farm labor.[2] Local yet extensive, the network of farm advisors throughout California and the United States provided a reserve of expertise that could be deployed when needed (Mukerji 1989). Although farm advisors' primary charge was to develop new and improved production practices, the war years highlight the flexibility of Cooperative Extension as an institution of repair. Among numerous wartime responsibilities, advisors in California became, essentially, labor contractors under the so-called Mexican National Program (later the Bracero Program),[3] which imported Mexican citizens to work in U.S. farm fields on a seasonal basis. These new responsibilities made advisors' mission more clear in some ways, less clear in others. The emphasis on increased agricultural production to aid the war effort focused advisors' attention on those factors that would most quickly stabilize and increase farm production; at the same time, farm labor was a very contested element of the farm system, and advisors struggled to balance competing interests.

To investigate these issues, I continue my focus on the Salinas Valley's vegetable industry but also include extensive discussion of the beet sugar industry, especially the case of the Spreckels Sugar Company. Spreckels was a beet sugar company that at one time operated the world's largest beet sugar processing plant on the outskirts of Salinas (in the area that is now actually a small town called Spreckels). Spreckels worked with government organizations on all levels to address its labor needs during World War II, importing Mexican labor under the Mexican National and Bracero programs and seeking to mechanize the production of sugar beets. In these endeavors, Spreckels worked closely with the UC and local farm advisors. The company's situation during and after the war illustrates my analysis of farm rationality, industry power, and repair because Spreckels "tried on" different constructions of crisis and repair at that time. In addition, the Spreckels case provides an interesting example of how different actors within the farm industry negotiated the meaning of the labor crisis. Because Spreckels relied on growers to grow sugar beets for its processing facility, it wanted these beet growers to perceive labor problems in the same way the company did and to accept the company's suggestions for repair. Unfortunately for Spreckels, growers often had their own agenda, and this balance of cooperation and conflict provides a very rich case for studying the negotiation surrounding problems, their definition, and frameworks for repair.

California's Growers and the Ongoing Farm Labor Problem

The labor practices of California's niche agricultural markets first took shape in the 1880s, when many growers in California began to abandon extensive crops like wheat and barley to cultivate intensive crops like fruits and nuts. Intensive crops often have much higher production costs than other crops, and labor costs are likely to make up half or more of this total cost (Taylor and Vasey 1936a; 1936b; Wells 1996). Thus, intensive cropping systems are sensitive to the cost of labor, and growers spent the first decades of niche market production experimenting with different methods of controlling the farm labor supply and labor costs. Perhaps the simplest way to keep wages depressed, from a purely economic perspective, is to have a labor supply that is much greater than demand. For the growers who pioneered California's niche farm industries, the basic demand for labor could be quite large, especially at harvest time. To keep wages low, many growers tried to recruit a workforce two or three times larger than basic demand required. Therefore, they defined a labor crisis not only in terms of the absolute number of workers required to complete the work but also as not having enough labor to keep wages low. Given these special demands, California growers seemed to face a labor crisis nearly every year, and "a labor shortage existed whenever [growers] were unable to hire sufficient workers to harvest crops at the utmost speed and at the most efficacious moment for the lowest possible wage."[4]

Growers found large supplies of labor by recruiting among poor migrant workers, especially recent immigrants. For example, when intensive agriculture in California first began in earnest, in the late 1870s and early 1880s, there was a large population of Chinese laborers in California, originally imported to work in mining and on the railroads. As employment in these two industries waned, agriculturalists began to use more Chinese labor. The use of "foreign" labor was initially controversial among the growers themselves, but they quickly realized the many advantages of employing recent immigrants (Daniel 1982, chs. 1, 2). Most of these workers had no land or permanent residence of their own, and their migrant status made it difficult for them to establish a long-term relationship with any one grower. Growers simply assumed that workers would appear for work during the busy planting and harvesting seasons, and then disappear to the cities (Chinese and Japanese labor) or back to their homeland (Mexican labor).[5] Growers also provided very poor quality housing (or none at all) to ensure that farmworkers would have little incentive to stay nearby at the end of each crop season (Guerin-Gonzales 1994, 62–63).

Growers' power over farm labor was supported in many ways by the state, from federal policies to local courts and police. When faced with the potential unionization of farm labor, California's agricultural interests called on the California State Highway Patrol, county-level sheriffs' offices, and local police to use violence and other forms of intimidation to suppress labor organization. These actions were often supported by county- and town-level antipicketing ordinances, with severe restrictions on rights of public assembly and protest. Although these antipicketing laws were eventually struck down by the Supreme Court of California in the early 1940s, local judges sympathetic to the farm industry initially upheld the laws, effectively making public protest illegal in many California counties during the second half of the 1930s.[6]

These local efforts were supported by racist state and federal immigration and citizenship policies that accorded recent immigrants few or no legal protections, drawing on and contributing to the larger discourses of racial hierarchy and biological determinism current in the late nineteenth and early twentieth centuries. These policies legitimated the exploitation of immigrant groups by stereotyping immigrants and their labor. As an example, a 1904 USDA report on the beet sugar industry emphasized the importance of getting labor for beet work that would not compromise American values, such as the importance of education:

In those old [European] countries the farmer and every member of his family were workers in the beet fields. Here the farmer and the hired men do the work; the children attend public schools and colleges. . . . It is evident that if sugar beets are to be produced in this country it must be done by a system of labor which is in harmony with American ideas, conditions, and aspirations. We could not adopt to any extent the family labor system prevalent in Europe. Our young people must not be deprived of educational advantages. (14–15)

But fortunately, in the eyes of this author, there existed in the United States groups of people who could be exempted from adopting these American values. These included recent European immigrants living in urban areas and other "foreigners" who were more suitable for this type of labor because they and their families could "secure immediate employment of the kind to which they were accustomed":

There is another class of foreigners, not previously experienced in growing beets, who readily adopted it on account of their natural adaptability to the system. As a class, they are accustomed to hard drudgery work of any kind, spending their lives during their stay in our country in work on public improvement, railroads, large contracts, etc., requiring hard manual labor. In this class come the Scandinavians (a few of whom have grown beets), Italians, Japanese, Chinamen, Portuguese, etc. Large numbers of these, annually increasing, take contracts in the beet fields. (37)

Racializing work and workers in this way provided a powerful cultural framework for legitimating inequalities.

These factors provided growers with a seemingly potent system for controlling labor costs and preventing the unionization of laborers. However, despite these advantages, the farm industry's control of labor was anything but certain, and two main factors contributed to the continual instability of growers' labor supply. First, the biggest and most conservative urban labor unions, especially the American Federation of Labor, feared that migrant farmworkers would quickly leave rural farming areas and move to larger cities, competing with white workers for urban industrial jobs.[7] These established unions used their political influence to push for the passage of anti-immigration laws, such as the Chinese Exclusion Act of 1882, which effectively shut the door on Chinese immigration and even provided a mechanism for many Chinese already in the United States to be repatriated. After California growers lost access to Chinese labor that matched their inflated demands, they were able to supplant them with Japanese immigrants, whose experience with orchard crops helped form the foundations of California's own nascent citrus industry. Just like the Chinese, however, Japanese farmworkers raised the ire of unions and white workers when the economy took another downturn, and Japanese immigration was also restricted (Daniel 1982, ch. 2). When the economy was good, growers found that they could again encourage another immigrant group toward the United States and California's farm industry. With each downturn, the immigrants were targeted and restricted again, and the cycle started anew.

Second, growers also faced a continual threat from smaller and more radical unions, which made an active commitment to organizing California's farmworkers in the late 1910s and throughout the 1930s. Provoked by poor working conditions and under the guidance of these unions, farmworkers belied their docile image by striking in a number of harvest seasons throughout the state. Several strikes in the Great Depression years, especially in the period 1931 through 1938, were particularly bitter struggles between California growers and a handful of small Communist party–affiliated unions. In some cases, these smaller unions scored successes against growers, mostly by timing strike actions for the harvest season of crops that needed to be picked quickly. These unions had more difficulty, though, building a permanent base membership from these actions, and growers typically responded to the unions' strikes and organization drives with violent strikebreaking tactics and legal action against the unions' leadership (Daniel 1982, ch. 7; Starr 1996, ch. 6).

Thus, at the onset of the Great Depression in the 1930s, the relationship between industrial agriculture and farm labor had a paradoxical quality, where stability and change were both present. Growers had been farming niche market crops for decades, and the industrial character of California's agriculture was relatively set by 1930. At the same time, growers and farm-workers were locked in a conflictual relationship that frequently brought protest and violence to California's farming communities. As the labor historian Ernesto Galarza notes, "The more [California agriculture] changed, the more it became the same thing" (1964, 107). One ethnic group was exiled only to be replaced by another; one union was destroyed and another rose to take its place. The constant was conflict, as growers struggled to maintain their control over labor relations, and workers and union organizers tried to change them. Each group saw the labor conditions of agricultural workers as a problem, but each defined the problem in different terms and proposed incompatible solutions.

For growers, the problem was often defined in terms of control. Tying into the anti-Communist movements of the 1930s, the farm industry blamed Communist "agitators" for disrupting the relationships that growers had established with particular groups of workers. Thus, labor problems were defined through factors deemed external to the farm industry and its labor relations. By framing labor conflict in this way, growers could portray the labor system as orderly, except for these external influences, and argue for its continuation. While, ostensibly, growers could have looked back at decades of conflict and envisioned a new way of organizing their relationships with farm labor, their actions repeatedly suggest a more conservative maintenance approach to labor relations. When faced with a direct challenge to their power, growers almost always adopted organizational strategies to preserve the existing labor system. For readers familiar with the history of factory- and mine-based labor conflicts during this same period, these strategies will sound very familiar. Beginning in the 1930s, the farm industry adopted organizational and public relations tactics that had been developed in conflicts between industrialists and factory and mine workers. This continuity was highlighted when the U.S. Senate investigated violations of U.S. workers' rights to free speech and assembly during strike actions of the late 1930s and early 1940s. Commonly called the La Follette

Committee, after Robert La Follette, a U.S. Senator from Wisconsin who led the investigation, the committee collected testimony related to several factory and mine strikes but also included evidence from some of the more prominent labor conflicts in California agriculture, emphasizing the continuity between stereotypically different industrial and agricultural forms of production and their labor relations (U.S. Senate 1940; Auerbach 1966). One of the farm labor conflicts that the La Follette Committee chose to investigate was a strike action by lettuce packing shed workers in Salinas in 1936.

The La Follette Committee's investigations of the 1936 strike reveal a complex, interlocking set of grower organizations that served to suppress dissent among the growers themselves, to win public support, and to undermine the unionization drive by the packing shed workers. Therefore, the key to growers' power and their ability to keep farmworkers largely unorganized over many decades lies in growers' own success at organization. Among the most important grower organizations for the 1936 packing shed strike was the Grower-Shipper Vegetable Association (GSVA). GSVA was (and remains) the keystone of grower organization in the Salinas Valley vegetable industry. Formed by a group of the most prominent lettuce growers in 1930, GSVA includes members from all parts of the vegetable industry: the growers who produce the crops, the packers and shippers whose companies ice and transport the vegetables, and the financial and agricultural input interests that support and profit from the industry. From its inception, a large part of GSVA's function has been to serve as a united front for growers to organize and resist potential labor conflicts, but the series of strikes in the early 1930s motivated growers across the state, including those in Monterey County's GSVA, to organize further and become even more vigilant in the face of labor activism. In 1934 growers and other interests affiliated with California agriculture created a statewide umbrella organization, the Associated Farmers.[8] Originally, the Associated Farmers acted mostly at the level of the state, holding conferences, influencing legislation, and organizing industry support to break labor actions. But as Roosevelt's New Deal labor programs, especially the National Labor Relations Act, threatened to give workers more rights to unionize, the Associated Farmers formed county-based chapters in major agricultural areas, including Monterey County. This Monterey County chapter of the Associated Farmers (MCAF) allowed the local vegetable industry to tap into the power of a well-funded and politically connected statewide organization. A third organization, the Citizens' Association of the Salinas Valley (CASV), was created as a local umbrella organization for other people and associations to support the growers' antiunion activities, especially the Monterey County Chamber of Commerce and the American Legion.

Although organization, through GSVA, had always been a key part of Salinas Valley vegetable growers' success, this tripartite grouping of GSVA, MCAF, and CASV allowed them to plan and organize in a way that consolidated local power and linked them to statewide sources of financial, organizational, and police power. For example, through affiliation with the larger Associated Farmers membership in California, Salinas Valley growers were connected to the money of financial institutions and the commensurate political influence to support antilabor bills in the California legislature. In addition, locally, GSVA's control over the vegetable industry allowed the most antiunion growers to pressure more sympathetic voices to resist labor organization at all costs. In the months prior to the strike, the minutes of GSVA's executive committee meetings show marked reluctance on the part of a subset of growers to resist unionization. Just before the strike began, in early September, growers were still divided over the best course of action to take. Some, like Bruce Church, one of the largest lettuce growers in the valley, favored negotiations with the union, admitting that wages should be raised slightly and that growers should not actively encourage a violent confrontation with the packing shed workers. Other members, however, such as Walter Farley of the Farley Fruit Company, were much more strident in their calls for resistance, claiming that the "issue is Communism" and that the growers should not "lie down before revolutionists."[9] Church's position remained a minority viewpoint, and the larger GSVA membership successfully forced dissenting voices to join the path of extreme resistance, using the threat of dismissal from the organization as a stick.[10]

The third organization, the Citizens' Association, also consolidated local power. CASV was intended to appear as a community-wide coalition of citizens supporting law and order as well as the common interests of employees and employers in Monterey County. Through CASV, growers hoped to shift the public's perception of the labor conflict from grower-versus-labor to a more dramatic and frightening scenario of a vulnerable small-town community holding off a Communist bid to control local business and politics. This CASV press release, which quotes CASV president Frank Cornell, an area tractor dealer, framed CASV as a wary representative of the "innocent bystander," a silent majority that had stood by for too long while Communist agitators took over a peaceful community:

What we are organized for is to keep the Salinas Valley on an even keel. We don't want anyone who has even the slightest stake in the community to get seasick.

We have no quarrel with any man—or group—who puts his full weight into the stroke of the oars and helps to shove the community ahead to the greater advancement of everyone.

But we haven't any sympathy, either, with the fellow who rocks the boat.

No one can speak conclusively the individual opinions of everyone in as large a group as this one is. But in general I know I can say definitely that certainly this association doesn't for an instant question the right of a man to organize and to act concertedly to attain definite objectives. How could we object? It's the very thing we are doing ourselves.

And I think we are doing the right thing.

For from a broad community point of view, the interests of employers and employees are absolutely mutual. And no minority group, whether they are classified as employees or employers, has any right by any stretch of the imagination to jeopardize the economic stability of a community which constitutes an overwhelming majority.

In the Salinas Valley, at least we hope that the "innocent bystander" is going to change his role. Instead of taking it on the nose in every brawl that happens, he's going to "walk softly and carry a big stick."[11]

This narrative of all community members' being in the same boat portrayed labor as a minority voice that could not understand (or had forgotten) the common interests between workers and management. By framing grower interests as allied with the interests of the larger community, the farm industry could justify the repression of this minority voice with its "big stick."

In all, Salinas Valley growers' access to resources for financial support, political leverage, media spin, and police-backed physical intimidation dwarfed the means of the packing shed workers; the workers went back to work after two months with no recognition or concessions from the growers.[12] In fact, the Salinas growers' plan had proven so successful at breaking the lettuce packers' strike that the same model was used in Stockton, in California's Central Valley, to break a strike of cannery workers in 1938.

This episode highlights the lengths to which California growers went to prevent the unionization of their workers and maintain tight control over farm labor. Growers tried to control the bodies of strikers (through intimidation and violence), the public's perception of the conflict (through public relations and CASV), and the solidarity of lettuce growers and packers (through pressure from GSVA). The sophistication of these practices is a testament to the importance of labor in California's niche market industries. Despite all this organization, planning, and action, however, labor relations in the lettuce industry stayed the same. Growers clearly saw the packing shed workers' attempt to gain union recognition as a crisis, a major threat to the structure of their industry, but their actions were aimed toward maintaining the status quo. Unionism was a problem to be addressed through intervention, and solutions such as bargaining and union recognition were ruled out as a matter of course. Those growers who entertained the idea of signing a contract with the shed workers were marginalized by others who used the ideology of anti-Communism and the interlocking power of the three associations to suppress dissent. In all, the growers deployed powerful cultural frames and organizational techniques to maintain their control and power.

Sympathetic and Critical Voices in California Agriculture's Farm Labor System

Aside from the financial, political, and police groups that California growers have relied upon to address labor conflicts, other organizations have provided support and advice to growers with respect to labor issues. For instance, UC scientists and administrators also recognized a farm labor problem and sought to provide counsel and solutions to help growers deal with labor issues. At the same time, others within the UC identified California agriculture's labor system as problematic, exploitative, and unsustainable, and challenged growers to make radical changes in it. These critics joined the activists who were directly involved in labor organizations by calling for an end to the status quo in California agriculture, presenting a powerful and very public counterpoint to growers' attempts to preserve and repair the labor system already in place. In this section I explore these divergent voices, their visions of labor use in California, and their solutions for its problems. As a large organization, the university accommodated a whole spectrum of practical and political positions with respect to the labor problem, and my discussion of these positions provides additional context to the story.[13]

Among the campus-based UC researchers and county farm advisors who were charged with improving California agriculture in the early part of the twentieth century, farm labor did not initially appear to be a central concern. Occasionally labor issues were addressed in passing, such as when the UC professor of agriculture Eugene Hilgard suggested in 1884 that cotton might be a good crop for California growers, not only because of its benefits to local soils but also because "by the spreading out of work over the entire twelve months cotton serves to secure steady employment, and therefore a steady laboring class" (Scheuring 1995, 33). It was not until U.S. involvement in World War I that the UC began to pay more attention to farm labor, owing to a labor shortage that foreshadowed similar problems during World War II. The UC assigned R. L. Adams, a member of the Giannini Foundation for agricultural economics on the Berkeley campus, to study and report on the farm labor situation in California. His reports, released in the late 1910s and early 1920s, became notorious among critics of California's farm labor system because of his racialized descriptions of farm laborers and the practical advice he lent to growers based on these distinctions. For instance, in an Experiment Station bulletin published in 1918, he claimed that "quarters provided for peon, coolie, or Oriental labor are generally not suitable for men demanding American standards of living" (Adams and Kelly 1918, 9). Further, in a textbook on general principles of farm management, Adams typologized racial and ethnic groups according to his perceptions of their characteristics, terming Mexicans, for example, in this way: "The common Mexican peon or laborer is usually a peaceful, somewhat childish, rather lazy, unambitious, fairly faithful person. He occasionally needs to be stirred up to get him to work, but if treated fairly he will work faithfully" (Adams 1921, 522; see also McWilliams 1939, 140).

In addition, Adams's typology included several other racialized and class-based stereotypes. These descriptions are crude, even by the standards of Adams's time, but the most interesting point about them is that they were intended as practical advice for growers to use in dealing with farm labor. Each subgroup within Adams's typology requires different treatment, and his recommendations for each group are tied to his perception of the specific characteristics and predilections of each. By manipulating the working and living conditions of these different ethnic groups and social classes, Adams maintained, growers could manage farm labor to better suit their production needs. With these instructions, Adams was not transforming California agriculture in any fundamental way but merely tweaking the current system to give growers more control and power over labor. Like the lettuce growers I described in the previous section, Adams framed labor problems and solutions as the maintenance of an established system of power.

In essence, this was the standard response of UC farming experts to the conflicts underlying California agriculture's labor system. Politically conservative and deeply connected to the farm industry, especially through the county farm bureau centers and the statewide California Farm Bureau Federation (CFBF), the UC College of Agriculture, the Experiment Station, and Cooperative Extension had little incentive to address labor issues with grand solutions. As the dramatic strikes of the early 1930s unfolded, the UC was relatively invisible during the actual conflicts. After a particularly bitter struggle between striking workers and growers in the Imperial Valley in 1933 and 1934, statewide farm organizations, including the CFBF, perceived that the strong-arm tactics used to break strikes were affecting the public's image of California growers. These groups asked California Governor C. C. Young to appoint a fact-finding committee to investigate the Imperial Valley strikes and to counter a report from a federal commission that was highly critical of the way local authorities and growers had handled the strike. The committee included the dean of the UC College of Agriculture, Claude Hutchison, who proved his loyalty to California growers and the CFBF by blaming the fierce temper of the strike on Communist agitators, not low wages, poor working conditions, or the repressive strikebreaking tactics employed by growers.[14] For Hutchison, then, California's farm labor system was problematic only with respect to the labor organizers.[15]

In contrast, others within and outside of the UC had much grander plans to change California agriculture and its system of farm labor to match their severe criticisms of California growers. These critical voices began attacking California agriculture in force beginning in the 1930s, when the most violent conflicts between growers and farm labor made headlines around the state and nation. Not coincidentally, this was also the period when the

Great Depression brought many thousands of homeless farm families to California from the Great Plains in the aftermath of the Dust Bowl. These mostly white families garnered much more attention from the media and from the critical voices than had the previous ethnic groups that worked on California farms.

Perhaps the most famous of the critics (aside from John Steinbeck) was Carey McWilliams, who published a number of articles sharply critical of farm labor practices in California throughout the 1930s, culminating with his book *Factories in the Field* in 1939 (also the year Steinbeck published *The Grapes of Wrath*). McWilliams's use of this imagery—a factory placed in the field—emphasized the strangeness of California agriculture vis-à-vis the U.S. ideal of family farming and drew on the (assumed) sharp divide between factory and farm production.[16] For McWilliams, the industrial form of farming in California, with large landholdings and capital-intensive production methods, was "irrational" at base and could not be repaired through modest change (1939, 22, 183). Ultimately, McWilliams believed, the only solution to the farm labor problem was a fundamental change in land ownership patterns: "Every solution [to the labor problem] which the growers have achieved has been a temporary solution, for the ultimate solution of the problem necessarily involves a basic change in the type of ownership and a breaking up of the large estates" (1939, 65).

McWilliams had no concrete plan for how this change could be implemented, although he spoke favorably of two experimental utopian farm communities funded through government agencies and premised on quasi-socialist methods of collective production and living. In contrast to his characterization of landholding patterns prevalent in California agriculture as irrational, McWilliams pointed to these communities as potential models for a different path based on rational state planning. Although both communities ended in failure, McWilliams placed the blame on incompetent government agencies and the ire of California business interests, claiming that "neither project was a fair test of what might be accomplished under state planning" (1939, 210).

Another critic was the UC Berkeley economist Paul Taylor, who published many academic articles on labor relations in California agriculture and served as a central expert witness in the La Follette Committee's hearings in California. Like McWilliams and other critics of the farm labor system in California, Taylor contrasted the industrial features of California agriculture with an idealized vision of family farming, and his testimony to the La Follette Committee hinged on this distinction. Taylor defined a family farm in terms of labor use, claiming that family farms used only the labor of their family members for farmwork, perhaps occasionally employing an extra "hired hand." He described California agriculture as a departure from this ideal and its labor practices as a deviation. In the following excerpt from Taylor's testimony, note how he uses framing words such as *common* and *abnormal* to portray California agriculture's use of farm labor as an aberration, leading to poor labor relations:

The recurrent conflict between employer and employee in the agricultural and processing industries of California, and in other neighboring states where similar conditions prevail, has been heralded widely as conflict between "embattled farmers" and "farm laborers." To describe the issues in these terms, however, is to mislead all who understand the words "farmer" and "farm laborer" as they are commonly used in other parts of the United States.

   The superintendent of the State Historical Society of Wisconsin, Dr. Joseph Schafer, recently stated clearly the traditional and well-understood meaning of "American farmer."

   "The farmer," he says, "is one who operates a 'family-sized farm for a living' rather than for 'an actual or potential fortune'; a farm on which the owner and his son or sons can perform the actual work of tillage, the female members of the household smoothing the way by providing home comforts, assisting about chores, or in field or meadow as pressure of work may dictate. Hired men are rather the exception than the rule in this typical agriculture. So far as they are employed, it is usually with the instinctive purpose of raising the labor force to the normal family plane rather than in hope of abnormally expanding the business beyond the family-farm size."

   The great strikes which periodically wrack the agricultural industry of California and may give rise to violations of civil liberties are not strikes between this kind of "American farmer" and his "hired man." In California, as in other parts of the country, the conspicuous instances of labor strife in agriculture occur between those individuals or corporations who are more properly called "agricultural employers" and the numerous workers who they employ for particular specialized operations such as picking, hoeing, or pruning during peak seasons at wages by the hour or the piece. There has been more strife in the agricultural history of California than elsewhere because here the number of farm employers who really are "agricultural employers" is so large, and because they, with their great number of employees, form an industrial pattern. (U.S. Senate 1940, 17215–17216)

Like McWilliams, Taylor used a distinction between the niche market industries and "family farming" to represent industrial agriculture as a kind of monster farming, a departure from the wholesome standards of the "American farm tradition." Taylor's solutions for the labor problems of California agriculture were slightly less utopian than McWilliams's, although just as likely to draw the wrath of California growers. Taylor attributed much of the labor problem to the seasonal demands for farm labor in the industry, which called for large numbers of workers during peak planting, thinning, and harvest times but only offered year-round employment to a much smaller group of permanent employees. Taylor suggested that growers coordinate their labor needs more closely, so that workers could maintain steadier and longer periods of employment each season. In this way, workers could also become more settled (Taylor and Vasey 1936b). Of course, from the growers' perspective, coordinating their labor needs with other growers would be an additional planning burden and would conflict with yearly variations in climatic and market conditions that affect growing schedules. Even more important, the casual way in which growers used farm labor accounted for a large part of the power that growers held over labor conditions and their ability to prevent farmworkers' unionization.

Both these sympathetic and critical voices had less practical impact on the actual form of labor relations during this period than did the growers and their extensive organizations, but each group contributed in important ways to the idea of farm labor as a kind of problem. For sympathizers like Adams, labor was a production problem to be managed, in order to help improve farm efficiency. For critics like McWilliams and Taylor, California agriculture had created a social problem through its system of farm labor, and generated "social consequences" that only radical change could address (McWilliams 1939, ch. 8). These different conceptions of labor as a problem pointed to two divergent paths for effecting change: repair as maintenance or repair as transformation. The two positions would resurface and blend in very surprising ways during the labor crisis of World War II.

Farm Advisors and Labor during World War II

As happened during World War I, when Cooperative Extension was quickly expanded and advisors were given broad powers to dictate the uses of farmland, encourage increased farm production, and regulate the use of farm labor (Danborn 1995, ch. 9; Scheuring 1995, 83–84), during World

War II advisors were given many new responsibilities for regulating the countryside and increasing farm production. The wartime work of UC Cooperative Extension included forming firefighting groups, mapping rural sources of water, encouraging and training homeowners to grow their own food in "victory gardens," and even forming a kind of rural militia for California countryside residents, where farm advisors were authorized to sign up recruits (Crocheron 1946, 7–8). Cooperative Extension served as a convenient point of access to rural communities and acted as a kind of organizational technology for the state, allowing the implementation of national policies on a broad but also very local scale.[17] Advisors, however, felt ambivalent about these new responsibilities. UC Director of Cooperative Extension B. H. Crocheron was already resentful of the increased regulatory duties placed on advisors during the 1930s as part of the New Deal.[18] After the end of World War II, he claimed that "no civil organization held a larger place in the war life of California" than Cooperative Extension, but he also struck a note of unease, terming the advisers' wartime activities a "rude interruption" (1946, 1). Of all the responsibilities that advisors took on during the war, their work in controlling the supply and placement of farm labor was perhaps the most onerous and with the most potential for conflict.

In 1939 and 1940, as formal U.S. entrance into World War II began to look more likely, growers began clamoring for permission to recruit labor in Mexico for agricultural work in California and other western states. Many growers probably recalled the labor conditions prevalent during World War I, when farmworkers were harder to find and commodity prices were sky high. During the first war, the United States and Mexico made an informal agreement whereby U.S. growers were allowed to travel to Mexico to recruit their own labor for agricultural work. This program was very popular with growers because it allowed them to recruit their own workforce, in numbers of their own choice, and with very few restrictions on the treatment and labor conditions of the Mexican workers. For the Mexican and U.S. governments, however, the legacy of this labor importation program was equivocal: growers had brought thousands of Mexican citizens into the United States for farmwork with little or no thought about these workers' fates when the war ended, U.S. soldiers returned home, and the demands of wartime production declined. As a result, in the years following World War I, the Mexican farmworkers faced a dire situation:

commodity prices and employment rates dropped, and labor unions and other groups began to protest the use of Mexican labor in industrial settings. Many employers, especially in unionized factory work, would not hire Mexican workers, and thousands of the workers recruited during the war were stranded in the United States with no hope for work or government relief. The United States and Mexico then cooperated again, this time on a program to send the workers back to Mexico, mostly at Mexico's expense. Thus, another episode in California's ongoing farm labor problem went through its familiar cycle of recruitment and exile, but this particular solution provided growers with a model for wartime labor shortages. Although the World War I recruitment program ended in controversy and great expense for the United States and Mexican governments, for growers it was a perfect solution, and the model was firmly in their minds at the onset of U.S. involvement in World War II (Reisler 1976, chs. 2, 3; Guerin-Gonzales 1994).

Despite the reluctance of both governments, and under intense lobbying pressure from farm interests, the United States and Mexico initially agreed to allow 1,500 Mexican citizens into the United States for fieldwork late in 1942. Of the initial 1,500 workers, 621 arrived by train in the Salinas Valley on October 5, 1942, primarily for work in sugar beet fields.[19] The following year, nearly 1,200 Mexican workers were sent to the Salinas Valley for farmwork. The USDA initially gave charge of the recruitment program, which at that time was generally referred to as the Mexican National Program, to the Farm Security Administration (FSA), an organization that had administered farmworker labor camps during the Great Depression years. FSA's involvement in the program was controversial from the start. Farm groups, especially the national American Farm Bureau Federation (AFBF), disliked FSA's past involvement with farm labor camps and opposed FSA plans to maximize use of domestic sources of labor before recruiting Mexican workers. AFBF continually pressured the USDA to take the Mexican National Program from FSA and to put it into the hands of the Cooperative Extension services in individual states. As AFBF's reasoning went, the Cooperative Extension services were meant to handle farm problems, and the labor crisis was a farm problem. Presumably, it also preferred to have farm advisors administering the program because, given the advisors' proximity to the growers, they could be more easily controlled. AFBF's pressure forced a compromise solution: FSA would recruit the

Mexican nationals, the state-level Cooperative Extension services would plan the interstate movement of domestic labor supplies, and county-based advisors in Cooperative Extension would facilitate the local placement of all forms of farm labor.[20]

In Monterey County the UC farm advisors were heavily involved in this kind of placement work, which included the formation of a Farm Production Council. These councils were established in each county and were a collection of important farm industry players. The councils not only allocated farm labor to specific farms but also had the power to set local wage rates for farm labor (Rasmussen 1951; Liss 1953). Fortunately, I was able to interview the retired farm advisor who worked most closely with growers on labor issues during the war and served on the Monterey County Farm Production Council. During our discussion, he emphasized the close connection between the advisors and the farm industry during this period, as they worked to place labor across Monterey County:

Retired Advisor: I think that in Monterey County one of the biggest fields that we got involved in was the better use of manpower. There was a shortage of labor because a lot of the fellows had gone off to war, so they started the Mexican National Program. While we were not involved in the actual working out of the details of the program, members of the Extension Service did work with growers, trying to help them understand the labor that they had and to use it more efficiently. One of the things that was organized in Monterey County, and I think we were a little different than most counties, we had what we called a Farm Production [Council].[21] As I remember there were fifteen members on this board, and the board was made up with growers from all over Monterey County, but primarily I would say that they were from the vegetable industry . . . and the sugar beet industry. They concerned themselves with the Mexican National Program. . . . In order to get a crew [of Mexican nationals] the [grower] had to be certified, based upon his needs. And the Farm Production Council became responsible for approving these requests for Mexican nationals. And this also helped with . . . the prisoners of war that were being brought in, and also the Farm Production Council approved requests for the use of the prisoners of war. The Extension Service was involved in that we had to go out in some instances and look at the field and talk to the grower and determine if actually these people were needed. They didn't want to have them approved and then find that when they get here they were not needed. . . .

Also, this Farm Production Council, which I happened to be the secretary of the thing . . . there were times when there were labor shortages in certain areas of the county, to perform certain jobs. And when that was recognized an effort was made to move the labor from, say, Salinas to Gonzales—let's say the labor was needed in Gonzales. And this council was very effective at doing that. It was one of the outstanding things I think that they did. And these meetings of this Farm Production Council were held every week at night, so it took a lot of time on the part of those that were involved and those of us that were in the Extension Service had to attend these meetings too.

CRH: How would you find out about, say, a certain need in a certain area, and how would you get the people down there?

Retired Advisor: Well, a lot of the big vegetable growers had labor camps and they had their own buses with which they transported the labor. And there also were some smaller growers that worked on a cooperative basis, that did the same thing. And then there were independent labor contractors who operated labor camps, and they had buses. And contact was made through these labor contractors and the [GSVA], which was an association of all the vegetable shippers, they acted also as a clearinghouse. [Growers] would phone [the GSVA and say,] "On Monday morning I want to start thinning, can you get me so many people to come and thin my field?" And so our activities were primarily to try to coordinate these activities and see that the labor was provided where it was needed.

From these comments, we can see the extent to which growers actually controlled the recruitment and placement of Mexican nationals and other sources of labor during the war. Although this retired advisor was an active member of the Farm Production Council, serving as its secretary, the council was dominated by representatives of the beet sugar industry (whose role I describe more fully in the next section) and the vegetable crop industry, through GSVA. These growers monitored labor needs, responded to requests for labor to be transferred to areas of the county with labor deficits, and even made the transfers in their own buses.

This close relationship, however, also had the potential for conflict. Like any other interaction with growers, advisors' involvement with labor issues meant negotiation, especially given that advisors were charged with stabilizing the farm labor supply. Growers and advisors did not always share the same meaning of *stable*, and the closeness with which advisors and the farm industry worked on such a sensitive problem was likely to create tension. This tension came out most clearly through the Farm Production Council's ability to set and adjust wage rates for agricultural labor throughout the county. The control the council held over farm wages was intended to stabilize the local farm workforce by eliminating competitive bidding for labor among growers, but it also allowed the industry-dominated council to keep wages artificially low through what was, in essence, collusion (Liss 1953).

Even before the war, UC Cooperative Extension lobbied growers to increase wages as a means of averting labor conflicts but the war years—and the tight control on wages held by the council—often put growers and advisors at odds over wage rates (Jelinek 1976, 210). For the local advisors, their involvement with this aspect of the council and its wartime activities was a tricky balance between their close ties to the farm industry and their responsibilities to stabilize the local farm labor supply. As one example, in a presentation given before a meeting of the Monterey County Farm Bureau, the assistant farm advisor Reuben Albaugh suggested that dairy farmers would have a more stable workforce if they simply raised wages:

Reuben Albaugh, assistant county agent, gave a very interesting report on dairy labor conditions, illustrating his talk with a chart showing the number of pounds of butterfat it took to pay various costs of producing milk for the past several years. This information indicated that dairy laborers are not being paid in proportion to the price dairymen are receiving for their product. This condition could be extended to other agricultural commodities. It was pointed out [to Albaugh] that raising wages of milkers was not the entire solution on the labor problem. (MCFB 1943)

It appears Albaugh received a tepid response during his presentation, but the wage rates set by the council remained an area of concern. By 1945 growers were becoming worried about the potential for labor unrest, given the controlled wage rates, and in 1946 the council reluctantly approved a wage increase to "avoid strikes and strong requests being made from labor" (MCFB 1945; 1946). During a subsequent meeting of the farm bureau, the merits of this increase were debated, and the farm advisor A. A. Tavernetti tried to make the bureau's directors see the wisdom in the decision:

A. A. Tavernetti explained the relationship of wages in comparison with net profit for ranch operations and brought out that usually it takes approximately one ton of a commodity to bear the expense of harvesting operations. Tavernetti further explained that with the present prices of commodities this wage rate increase was not out of line. . . . After a thorough discussion of the matter, those in attendance found that the Farm Production Council had made a wise decision and that such action was necessary to avert further trouble. (MCFB 1946, 1)

Despite the potential for conflict between growers and the state (including Cooperative Extension), it would be a mistake to represent this tension as a broader conflict between incompatible visions of repair. Clearly, farm advisors were somewhat reluctant to get entangled in the politics of farm labor in California, and yet it was growers themselves who lobbied for Cooperative Extension's involvement in the Mexican National Program. Further, the very fact that farm advisors dared to go before farm groups during this time period and seriously suggest wage increases shows the extent to which advisors were integrated with the local farm community and could suggest controversial, if modest, forms of repair. In the end, advisors' suggestions for small wage increases still represented a maintenance approach to repair, and in many ways the Mexican National Program was the ultimate maintenance plan for the growers' system of farm labor; it repaired the system while making few or no concessions on growers' power. The use of farm labor during the war years was essentially under direct grower control, but with extensive support from all levels of the state. For all these reasons, growers viewed the Mexican National Program as a great success, and niche crop industries across the State became more and more dependent on the use of Mexican national labor in this period.

Maintenance or Transformation? The Spreckels Case

One farm sector in California that became very dependent on Mexican national labor was the beet sugar industry. Unlike with many vegetable and other niche market crops, however, the use of this source of labor during and after the war years created a tension between the use of hand labor and plans for a more transformative change: fully mechanized production of sugar beets. In this section I describe how one company, the Spreckels Sugar Company, struggled to reconcile these two modes of repair in the Salinas Valley.

Why treat the case of Spreckels in such depth? Overall, the Spreckels case illustrates the connections and tensions between industrial modes of farm production, labor politics, and the role of the state during this period. Labor was a continual problem for the company's operations in the Salinas Valley, and these labor troubles were not just a sudden outgrowth of World War II. The sugar beet industry's labor problems stemmed from the basic constraints of industrial production and the local political ecology of Salinas Valley farming. Sugar beets may have been California's first truly industrial crop because they are not eaten as beets—they are produced solely to be processed into sugar in a processing factory. Therefore, sugar beet farming is tied very closely to the constraints and needs of industrial processing, and processors require a steady and orderly flow of beets to keep their factories operating at peak efficiency. As a consequence, beet processing was heavily reliant on labor—for operating the factory, of course, but also to plant, thin, and harvest beets on the thousands of acres required to support a large processing plant. The war tightened these constraints and led Spreckels to become one of the most aggressive lobbyists for the Mexican National Program and among the largest recipients of Mexican national labor during and after the war.

The interplay of these factors can be traced back to a dramatic event in the history of the Salinas Valley: in 1898 it became home to what was at the time the world's largest beet-processing facility. The United States had no real sugar beet industry prior to the 1890s, but subsequent decades saw a marked increase in beet sugar production in response to a federal system of incentives.[22] Built by Claus Spreckels, the patriarch of a sugar empire throughout the western United States,[23] the plant was intended to take over a large portion of the valley's farmland for beet production, as emphasized in a USDA report of the time:

One of the remarkable incidents connected with the beet-sugar industry in this country during the past year was the building of that mammoth concern at Salinas, by Mr. Claus Spreckels, of California. This gives the United States the largest factory in the world. . . . The factory will cost $2,750,000. It will use about 1,200 barrels of petroleum daily for fuel. . . . Its production of raw sugar will be about 400 tons per day. It will require about 30,000 acres of ground to produce the beets. (USDA 1898, 16; Allen 1934, 44–45)

The 30,000 acres of beets required to operate the plant represented about 15 percent of the valley's farmland, so Spreckels needed to convince many growers to sign annual contracts for the production of sugar beets, which set a price in advance for beets delivered to the company at the end of the season. Ideally, Spreckels would sign enough contracts with growers each year to provide enough beets for the factory to operate at full capacity for about 120 days, generally from the end of August until mid-December (Pioda, *History*). In practice, however, the company struggled to get enough beets from the very start.

The year of the Salinas factory's completion, 1898, was a drought year, and only 7,200 acres of beets were harvested. For this first year, Spreckels decided not to open the Salinas factory but to ship beets to the Watsonville factory for processing (Pioda, *History*). This false start was just the first of many problems for the nascent sugar beet industry in the Salinas Valley. Perhaps the most important, at first, was the pattern of rainfall. Rainfall around Salinas averages from 10 to 15 inches per year, most of which falls from November to April; during the summer months, little rain falls. Sugar beet growers found that conditions were too dry through the summer for a good crop, and yields varied considerably from year to year, contingent on the amount of rain. Also, a disease of sugar beets called beet blight was often prevalent and did great damage to each year's crop. By 1903 the company decided to abandon beet farming on one tenant ranch simply because beet blight was prevalent there every year (Pioda, *History*).

Spreckels's difficulties in getting beets and keeping the factory in production are illustrated in figure 4.1, which depicts the yearly harvest of beets and days of factory operation for the first few decades of the twentieth century.[24] Prior to 1914 beet acreage in the valley never exceeded 20,000 acres, far from the 30,000 acres mentioned in the 1898 USDA report. Until 1918 the number of days the factory was open to process beets tracked the number of acres planted in the valley quite closely, and in many years the factory was only open for 60–80 days. Starting in the late 1910s and through the 1920s, Spreckels was able to ship beets from other growing areas to process at the Salinas plant, and this accounts for more stability in the number of plant operating days. In fact, even when the acreage of sugar beets grown in the valley virtually disappeared in the years 1928 through 1931, because of a major shift to vegetable crops, the plant was still able to operate for about 100 days each year. However, this meant that the Salinas factory processed beets at the expense of other Spreckels plants.[25]

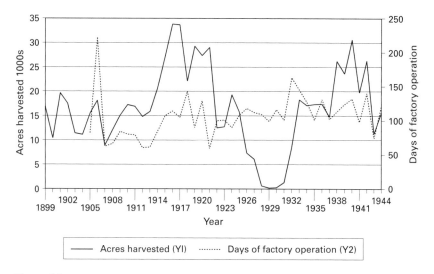

Figure 4.1
Beet acreage and days of factory operation, Spreckels factory, Salinas, California, 1899–1944. From Pioda, *Chronological History, Spreckels Sugar Company.*

Even with beets shipped in from other areas, the Salinas plant worked well below its full capacity most years. Innovations in irrigation and disease control in part account for the relative success of the plant from 1915 to 1921. But the mid-1920s brought yet another challenge: the rise of the vegetable industry in the valley. At the same time that Spreckels was finally beginning to make fuller use of the mammoth Salinas facility, local growers were beginning to experiment with vegetable crops, especially lettuce. Increasing throughout the 1920s, lettuce production claimed 50,000 acres of valley farmland by 1930, a time when sugar beet produc-tion was virtually nonexistent. Charles Pioda, the longtime superintend-ent of the Salinas facility, lamented the competition from vegetable crops in 1924: "Artichokes, lettuce, and other garden crops [are] making strong inroads in our beet territory," and he noted that in 1929 none of the tenant growers who rented company land were growing any beets at all (Pioda, *History*).

When the onset of the Great Depression caused commodity prices for vegetable crops and dry beans to drop dramatically, many growers sought refuge in sugar beets, leading to a rise in beet acreage throughout the 1930s. Even though Spreckels was paying less per ton in grower contracts than it had in more than a decade, production of beets quickly passed 15,000 acres in 1932 and 25,000 acres in 1938 (see figure 4.1). This movement back and forth between vegetables, sugar beets, and other crops points to a complex calculus of annual decision making for growers in the valley. Sugar beets were just one crop that growers could choose to plant. The decision to grow a certain crop was based on a multitude of interrelated factors, including experience, financing, soil conditions, access to water, and the overall capital intensiveness of the crop. Among all these factors, one overriding concern governed a grower's crop choice: which crop would bring the greatest monetary return. Unfortunately for Spreckels, this calculus often went against sugar beets and in favor of other crops. The growth of vegetable farming, especially lettuce, encouraged a kind of speculative agriculture, much like the "mining" style of wheat farming begun in the 1860s when wheat prices were especially favorable. Despite the fact that lettuce is difficult to grow and requires a large investment just to bring a crop to market, it can be one of the most profitable crops of all when its price is high. This "green gold" seemed especially lucrative throughout the 1920s because the demand and price for vegetable crops were very high.

The movement back to sugar beet production in the 1930s pointed to one advantage that sugar beet production had over vegetable production: price stability. Because Spreckels usually offered growers a set price for beets when they signed a contract at the beginning of a crop season, growers could count on a minimum level of return from their investment, given good climatic conditions and the successful growth of a beet crop. Therefore, when commodity prices for vegetables were low or expected to be low, growers often turned to sugar beets. When prices for vegetables were high, they moved again to produce "green gold." This movement between vegetable crops and sugar beets points to the evenness of other basic production factors that growers considered when comparing vegetable and sugar beet growing. For instance, both vegetables and beets require irrigated land, pest control, and large amounts of hand labor to produce a crop.

One way for Spreckels to make sugar beet farming more appealing was to change the factors that were the same for both beets and vegetables so that one of them would go in favor of beet production. Water needs were non-negotiable, as were requirements for good land; both vegetables and beets did better on good land that was irrigated.[26] Similarly, the prices of vegetable crops shaped the land valuations in the Salinas Valley, making it more expensive to rent land for any kind of crop. If a grower chose to move onto less expensive land, it would mean poorer soils, less desirable topography, and a poorer crop. Labor needs were also high among both crops; each required intensive handwork for thinning, weeding, and harvesting. Labor was the greatest outlay that a grower made when bringing either sugar beets or vegetable crops to harvest, usually representing more than half of a grower's total production expenses (Allen 1934). Therefore, if Spreckels could find some way to make sugar beet production less labor-intensive, this factor could be shifted in favor of beet farming.

The Drive for Fully Mechanized Sugar Beet Production

Given Spreckels's beet supply difficulties in Salinas, U.S. involvement in World War II held mixed prospects for sugar production. The potential for a real turnaround in Spreckels's luck was certainly there. Demand for sugar from domestic sources surged during the war years as supplies of sugar from cane produced in the Pacific fell under Japanese control and became unavailable to the United States and its allies. As the advertisement in figure 4.2 boasts, beet sugar was the "sugar no enemy can touch" (assuming the Axis forces were kept off U.S. soil). Sugar was a war commodity, not only for consumption by U.S. and Allied soldiers but also, when transformed into industrial alcohol, for the manufacture of explosives, synthetic rubber, and plastics (figure 4.3). The war also affected the sugar supply of civilians in the United States and Allied countries. U.S. sugar manufacturers were suddenly responsible for supplying sugar to many millions more people than before the war (figure 4.4). All these factors together created a high demand for sugar and a virtually limitless market. Given that the sugar market had been relatively flat for several years prior to U.S. involvement in the war, 1941 should have been the start of a banner period of profits for Spreckels.

Once again, labor problems thwarted these chances, and Spreckels could not get enough beets to operate at full capacity for a good part of 1942 and 1943 (see figure 4.1).[27] These early war years saw a massive scramble to mobilize new sources of labor and new production methods to increase output for the war.[28] Among the strongest advocates for the Mexican National Program was a group called California Field Crops, a lobbying

Here is SUGAR no enemy can touch

YOU'RE LOOKING AT a sugar beet. With Philippine sugar cut off, with a need to send sugar to our allies and perhaps also turn much sugar into gunpowder, the sugar beet has become America's most important single source of sugar.

Last year the sugar beet furnished one-quarter of all the sugar consumed in the U. S. Last year 65,000 western farmers grew this vital crop.

This year sugar beet acreage is being increased. The industry and the government are working night and day to make this increase as big as possible.

*   *   *

In the meantime, it is necessary that sugar be rationed. Here are the reasons:

● Because increases in beet sugar cannot as yet make up the loss of Philippine and other cane sugar.

● Because America must supply its allies with sugar in addition to supplying our own people.

● Because factories can utilize sugar to make the industrial alcohol with which smokeless powder is manufactured. In order to get enough explosives *quick*, we may have to turn over 1,000,000 tons of sugar into alcohol. Some sugar is being used for that purpose right now.

*   *   *

Let's remember several things as we cut down on sugar to the ration level. First, let's remember the cup-or-so of sugar we save each week by rationing may, in the form of gunpowder, save an American boy's life.

Second, let's remember that we are not being asked to give up much—the proposed sugar ration is larger than the one we took in stride back in 1918.

And lastly, let's remember that while we are cutting down on sugar under the ration, we will never have to *go without* sugar. Because we now grow sugar here *inside* our own country—in the form of billions of sugar beets like the one pictured at the right.

The sugar beet is long and tapering, silver-white in color. It is different from the small red table beet often grown in home gardens

The largest-selling sugar

grown in the West

Figure 4.2

Spreckels advertisement promoting the advantages of a home-grown sugar supply during World War II. From Spreckels (1942a). Courtesy of Monterey County Department of Parks.

SUGAR explodes

Bombs, made partly from sugar, are blasting our enemies on all fronts

Every time an American pilot drops bombs on an enemy target, some of America's normal civilian supply of sugar is sent hurtling down to explode.

That's because it takes industrial alcohol to make bombs, and it takes sugar to make industrial alcohol.

So much alcohol was needed for war that an estimated 800,000 tons of sugar was diverted from food purposes and put to work as war material in 1944.

SUGAR makes Tires

How to keep 'em rolling? An almost unanswerable question in those first weeks after Pearl Harbor, with our usual source of rubber cut off. But the answer was found in sugar.

Synthetic rubber is one of the most important products derived from industrial alcohol — made largely from sugar. During the coming year, industrial alcohol will supply to the synthetic rubber industry about 53 percent of all the butadiene — the chemical element used to make synthetic rubber.

An army no longer marches on its stomach, but literally on sugar, from which comes a vital ingredient for synthetic rubber

Thus sugar helps provide the tires to keep our armies on the move — and supply the other required items formerly made from rubber.

Sugar has not only gone to war, but is transporting it!

SUGAR in Plastics

Sugar goes into plastics, such as this "greenhouse" for the boys upstairs

Sugar in the form of industrial alcohol is pouring into war plants of industry.

And the manufacture of plastics by the chemical industry is one of the important results; plastics as used in "greenhouses" for planes, and for an increasing list of war commodities where plastics are substituting for metal!

Because of the unusual and increased demands of war, industry found itself dependent on its research departments to find substitutes for materials no longer obtainable, and to provide new materials to meet new demands.

Sugar, in playing a vital part in chemical research, has not only gone to war but will have much to contribute to the products of peace.

SUGAR Outlook

When manpower demands from all sorts of war industries dried up the usual labor supply, sugar beet growers were faced with finding ways of harvesting their crop with fewer workers.

One answer to the problem was the beet harvester, many of which were purchased by the Spreckels Sugar Company for rental to growers.

This mechanical beet harvester lifts beets from soil, tops and loads beets on truck. It offers great promise for the complete mechanization of sugar beet harvest

In field tests it was found that this harvester delivered beets in the truck at less cost and greatly reduced labor requirements. So the outlook for the farmer who raises beets is brightened, thus brightening the outlook for an increased supply of sugar as sheared seed and mechanical harvesters come into wider use.

Figure 4.3
Spreckels advertisement explaining the use of sugar for manufacturing in the war industries. From Spreckels (1945). Courtesy of Monterey County Department of Parks.

proxy of four beet sugar processing companies, including Spreckels. The group requested 3,000 Mexican workers for the 1942 sugar beet harvest, and Charles Pioda, the superintendent of the Salinas factory, traveled to Mexico with FSA officials to oversee the selection of the first workers. The early results of the program were mixed. While some of the workers were experienced in farm labor, others were not, and few workers had experience with sugar beets. Further, since most sugar beet growers were accustomed to contracting out their fieldwork to labor contractors, few growers had much experience in educating workers on how to do beet work. The UC attempted to fill this gap with "emergency farm labor leaflets" such as the one in figure 4.5, which demonstrates fieldwork techniques to improve

How many babies to a buggy?

S-T-R-E-T-C-H

WHEN you're expecting a single...and *three* show up ...you're mighty proud. But it means stretching bonnets and blankets and things.

It's the same with the sugar situation. Demands are still heavy and it's quite a task to make the supply go around. So, sugar rationing will probably be with us for some time to come.

War-ravaged fields abroad must be restored before we can increase the tonnage in our sugar barrel. It can easily be a year or more before world sugar supplies are ample!

That's the sugar problem. We at Spreckels are keeping our factories working day and night to produce the sugar our Western consumers need.

S-T-R-E-T-C-H your supply of this home-grown sugar as far as possible. It's pure sugar from Western farms.

SPRECKELS SUGAR

HONEY DEW

SPRECKELS SUGAR
HOME-GROWN IN THE WEST

Figure 4.4
Spreckels advertisement emphasizing the heavy demands for sugar to be met by U.S. beet growers. From Spreckels (1946a). Courtesy of Monterey County Department of Parks.

workers' efficiency. Note how the leaflet emphasizes the specific bodily motions and procedures for most efficiently harvesting beets by hand (See also Pioda 1943.)

As they gained experience, the Mexican workers became more acceptable to growers, and Spreckels's beet growers (and the valley's vegetable growers) became increasingly reliant on the Mexican National Program for their labor needs. Thus, Pioda was able to write of the 1945 Spreckels campaign in Salinas,

There was no serious shortage of field labor in 1945 in this district. There were times when more labor could have been conveniently used, but no beets were thinned or hoed so late that crop damage resulted from scarcity of labor. The available labor

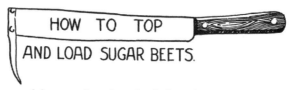

HOW TO TOP AND LOAD SUGAR BEETS.

Very successful systems have been worked out for harvesting sugar beets by hand. Crews handling the largest tonnage with the least effort have developed a method which does away with all useless motions. The following outline shows one of the best methods.

The beets should be plowed in units of 16 rows from one direction. The toppers should work facing in a direction opposite to the way the plow traveled. Due to the slight tilt the plow has given, the beets will lift out easily.

Eight men work as a topping crew. Use the standard beet knife with the hook. Keep the knife sharp and the hook pointed.

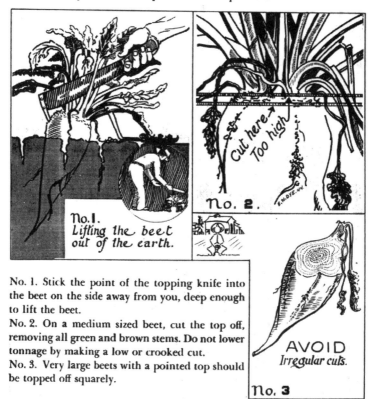

No. 1. Stick the point of the topping knife into the beet on the side away from you, deep enough to lift the beet.

No. 2. On a medium sized beet, cut the top off, removing all green and brown stems. Do not lower tonnage by making a low or crooked cut.

No. 3. Very large beets with a pointed top should be topped off squarely.

Figure 4.5

An "Emergency Farm Labor Leaflet" published by UC Cooperative Extension in 1944. Courtesy of UC Department of Agriculture and Natural Resources.

at harvest time and the harvest machinery not only harvested the crop as fast as the plant could mill the beets, but provided so great a daily surplus that the harvest had to be regulated. (*History*, 9)

This source of labor became so popular among beet growers that it undermined the second major strategy that Spreckels and the UC used to address the labor crisis: mechanization of beet production.

At the onset of U.S. involvement in World War II, it did not take a great leap of imagination to envision the mechanized production of sugar beets and realize the advantages that such a system would give Spreckels in its constant quest to increase beet production. Mechanization had the potential to shift the balance between sugar beet and vegetable production in the Salinas Valley by decreasing the costs and conflicts associated with farm labor for beet growers. Steps toward mechanized production were already being taken several years before Pearl Harbor. Informal cooperation on mechanization between the sugar beet industry and the UC began as early as 1935, with a more formal agreement between the university's Board of Regents and a committee of industry representatives in 1938 providing $70,000 for research over a three-year period (Spreckels 1938). Through the late 1930s and into the early 1940s, Spreckels updated growers on the progress and promise of fully mechanized sugar beet production through its *Spreckels Sugar Beet Bulletin*.[29] In a 1939 issue, Monterey County farm advisor A. A. Tavernetti reported on field trials conducted in the Salinas Valley, trials that tested mechanized methods of thinning sugar beets. Traditionally, sugar beet growers hired field-workers for intensive work at two times of the year: in the spring for thinning and weeding nascent beets, and in the fall for harvesting them. Workers thinned the beets by hoeing out extra beet plants and weeds just after they emerged from the ground, leaving an evenly spaced number of single beets in each row. When the beets were spaced in this way, growers were assured that individual plants would not compete with each other for nutrients and water and that each plant would grow as large as possible before the harvest. Even more important, a row of larger, neatly spaced beets could be hand-harvested with beet knives easily and quickly (and therefore cheaply). Thus, this beet-thinning technique was a long-standing practice in the sugar beet industry.

In Tavernetti's experiments, though, the principle of the "single"—a solitary beet growing without competition from other beets in the same

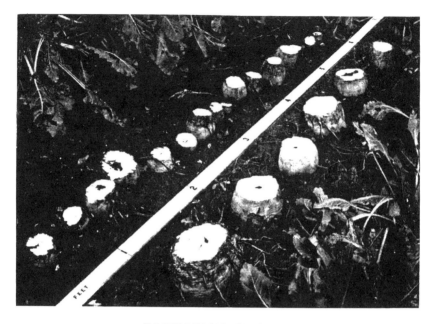

TAKE YOUR CHOICE

The two extremes pictured above yielded the same sugar per acre.

THINNING

WEED CONTROL

HARVESTING

Under a mechanized plan can be profitably accomplished with
a wide tolerance of spacing in the row.

Figure 4.6
Photo from *Spreckels Sugar Beet Bulletin* showing machine-blocked sugar beets (*top*)
and hand-thinned sugar beets (*bottom*). From Spreckels (1953). Courtesy of Monterey
County Department of Parks.

row—was violated; instead a blocking machine was used to simply cut a
single row of plants regardless of spacing. Figure 4.6 makes these competing strategies clear, depicting in the bottom half a traditionally thinned
row of larger singles and in the top half a machine-blocked row of smaller
but more numerous beets. In his report Tavernetti describes the distinction
between the two methods and the changes in practice required for mechanized production to work properly. Then he comments,

It is quite obvious no machine can be perfected that can duplicate hand work as is
now being employed in thinning sugar beets.

If sugar beet growing and harvesting is completely mechanized, the probability is that it will be due to the fact that certain details now employed in hand work are not indispensable or that sufficient savings in money can be effected to offset some losses in ultimate sugar per acre. (1939, 1)

Thus, the crucial test for machine thinning lay in whether it could produce comparable results at a cost savings over hand thinning. Tavernetti reports some promising figures from his field trials, showing that the weight of sugar extracted from an acre of machine-blocked beets was essentially the same as that taken from an acre of hand-thinned beets; in other words, more small beets were equivalent to fewer large ones. However, he ends with a cautious note: "There are a large number of other problems involved which must be studied and understood before the grower can afford to take the risk of eliminating all hand thinning" (2).

The onset of the war two years later, the depletion of domestic labor sources, and the initial reluctance of the United States and Mexican governments to begin a new labor recruitment program pushed mechanization to the foreground. Although the technology was still experimental, Spreckels and UC farm advisors began encouraging beet growers to consider mechanization to grow and harvest their crops. Spreckels hired Austin Armer, a specialist in agricultural engineering, to adapt and improve work on sugar beet mechanization that he had begun with the USDA (Spreckels 1943). Spreckels also purchased thinning and harvesting prototypes of farm machinery for Armer and his staff to experiment with and to field test with local growers.[30] Throughout 1943 the *Spreckels Sugar Beet Bulletin* featured reports on these initial trials, with examples of local growers' successful use of the machinery for thinning and harvesting expressed in bold headlines such as "COSTS CUT WITH DIXIE THINNER" (Lambdin 1943; Crane 1943). The following year Armer (1944) was encouraging growers to carefully plan their crop plantings for the 1944 season in order to take advantage of the advances made in the sugar beet harvesting technology. Spreckels also made several harvesters available for grower use during the 1944 campaign (Spreckels 1944).

The initial use of the new machinery for the spring and fall beet work did not live up to the hype, and few growers showed interest in mechanized production. Monterey County farm advisor A. A. Tavernetti (1943) also continued to strike a more balanced tone, informing growers through the *Bulletin* that thinning costs had risen dramatically during the war years but commodity prices had kept pace, so that growers were still paying about the same proportion of their yield toward labor costs. For those who did try machine harvesting during the 1944 season, the results were mixed. As Pioda noted in his *Chronological History*, the harvesting machinery showed promise, but good results were highly dependent on proper planning by growers and good field conditions:

Previous to the start of the harvest, the Company purchased six Marion Harvesters for use by growers in this district. These harvesters were rented to growers at the rate of $7.50 per acre. Two additional machines were purchased and used by growers. About 25,000 tons of beets were harvested by machine. Working under extremely favorable conditions, these machines did an acceptable job of topping and loading under present abnormal labor conditions and subnormal standards of acceptable work. Topping was badly done and the breakage of roots was high. There were a great many beets left in the fields. One grower at King City, whose field conditions were ideal and who maintained a well-equipped machine and repair shop, harvested 1400 tons in 10 days and had a top output of 208 tons in one ten-hour day.

Despite these decent results under the proper conditions, the reliability of the harvesting machinery was not nearly as promising:

The machines were not in operation 50% of the time. Replacement parts were purchased and installed but either broke or wore out shortly after installation. Growers became thoroughly disgusted with the poor performance of the harvesters and finally refused to attempt to make use of them.

Spreckels nevertheless kept pushing growers to adopt mechanized harvesting and made its own modifications to the harvesting machinery to improve its robustness and overall applicability to beet production. Local farm advisors also gave detailed technical advice to growers on the plowing, planting, and other initial crop conditions that would mesh best with a mechanized harvest at the end of the season (Tavernetti 1946). Beginning with the 1946 season, many growers reconsidered the advantages of mechanized fall work and began a major shift toward the mechanical harvesting technology. Figure 4.7 shows a cover of the *Bulletin* from 1946 titled "Mechanical Harvest Increases Grower Profits" and giving cost figures to justify growers' use of the harvesting machinery. Another issue claimed that 25,000 acres of California sugar beets would be machine-harvested in 1946 (Spreckels 1946c). Spreckels's boast was justified in this case: the 1946 season saw 28,469 acres harvested by machine (Armer 1947).

PUBLISHED FOR CALIFORNIA SUGAR BEET GROWERS BY THE SPRECKELS SUGAR COMPANY

Vol. X   JANUARY-FEBRUARY 1946   No. 1

A typical view of the two-row Marbeet harvester in operation. The loaded truck can be replaced by the empty one without loss of harvester time as the loading of beets can be stopped by disengaging the beet elevator clutch without stopping the harvester. The efficiency of operation of these machines depends to a large degree upon the attention paid to management detail such as the one illustrated in the above picture.

MECHANICAL HARVEST INCREASES GROWER PROFITS

*COST PER TON—HAND HARVEST $2.13
*COST PER TON—MACHINE HARVEST........ 1.28
NET GAIN TO GROWER............................**$ .85**

Figure 4.7
Cover from *Spreckels Sugar Beet Bulletin* picturing a mechanical sugar beet harvester loading beets into trucks. From Spreckels (1946b). Courtesy of Monterey County Department of Parks.

In contrast to the success of the fall harvesting mechanization, the spring thinning work was much harder to mechanize, both with respect to the engineering involved and growers' preference for hand labor. Growers proved reluctant to adopt the new blocking technology for spring work, as Pioda noted in *Chronological History*,

No progress was made in 1944 in pre-harvest mechanization. Growers will not use either the Dixie or the Farmer Mercantile Company blockers so long as hand labor is available. Growers insist upon a regularly spaced stand of beets. They are not convinced that they can secure a satisfactory stand from mechanical blocking. Because of the shortage of hand labor and resulting high wages paid to workers, there has been no reduction in costs following mechanical blocking.

Pioda makes a very similar note regarding machine thinning in his 1945 installment of the *Chronicle*. In 1946, when the *Bulletin* crowed in every issue about the success of mechanized harvesting, there is virtually no mention of mechanized thinning. Many problems made growers more hesitant to mechanize the spring thinning work than the fall harvesting work. First, if a grower wanted to harvest with hand labor, the smaller, more numerous beets resulting from mechanized thinning would take much longer to harvest and cost more. Second, the beets needed to be weeded at least once per season, and this required hand labor anyway, so why not have them hand-thinned as well? Third, because of the method by which the mechanical thinner worked, growers needed to use more seed per acre to ensure that there were not gaps in a given row.

These factors worked against the use of the new mechanical thinners, but, ironically, it was the beet sugar industry's initial solution to the war labor crisis, the Mexican National Program, that impeded the thinning technology most of all. As noted, after troubles early in the war, the seasons of 1944 and 1945 had few problems with labor shortages, and given the unreliability of the thinning machinery and its special requirements, growers were reluctant to mechanize as long as there was an adequate supply of cheap labor through the Mexican National Program. These preferences did not change when the war ended. In fact, the Mexican National Program proved so popular with growers that they lobbied for its continuation well past the end of the war. Spreckels was caught between its desire for fully mechanized beet production and its need to retain the beet growers who had become dependent on imported labor.

The Perpetual Crisis: Labor and Mechanization in California after World War II

Although Spreckels and other beet sugar processors were key players when the Mexican National Program was begun in 1942 and hosted the first Mexican workers, as the war years progressed, more and more niche market industries began using Mexican farm laborers, and growers across California quickly came to rely on them as a source of farm labor. When the war ended, growers did not want to give them up. At first, agricultural groups were able to justify the continuation of the program by pointing to the devastated economies of Europe and their lack of native food production; later, growers simply argued that their industries would not survive without imported labor. The years just following the war saw several informal agreements between the United States and Mexico to renew the recruitment program on a yearly basis until Public Law 78, federal legislation passed in 1951, formalized the two countries' agreements. Like the Mexican National Program, the Bracero Program under Public Law 78 contained stipulations for growers' employment of Mexican workers, including requirements for a fair prevailing wage, transportation to and from work sites, and protection from discrimination. Bracero labor was also supposed to be used only when domestic labor was not available. However, most of these requirements were ignored by the Farm Placement Service, a division of the U.S. Department of Labor that administered the recruitment and placement of bracero labor and essentially acted as a government-subsidized farm labor contractor for the farm industry in California and the Southwest.[31]

The use of bracero labor represented the best of all possible worlds for California growers. Braceros went straight back to Mexico after their seasonal contract ended, in contrast with domestic workers, who tended to settle in California's agricultural valleys. Owing to the transitory character of the bracero workforce, organizing these workers was all but impossible. Also, braceros were mostly young single men; growers didn't have to worry about where the braceros' families would live or what they would eat. This also made it quite easy for growers to use the threat of deportation to defuse any protest. Finally, braceros could be imported in huge numbers to flood the labor market and keep wages very low compared to other industries.[32] As an example of this last advantage, Salinas Valley lettuce growers were able to settle their long-standing labor dispute with packing shed workers by moving much of the packing work into the field, where it was performed instead by bracero labor (Galarza 1964; 1977). Given all these factors, access to bracero labor not only maintained growers' control over farm labor but actually consolidated and increased it.

For growers, the system was almost too good to be true, and they fought for the Bracero Program's continuation throughout the 1950s and early 1960s, as mounting pressure from union and civil rights groups threatened to end the program. Thus, agricultural groups complained of an impending labor crisis nearly every year, hoping to extend the use of bracero labor as long as possible. Given the huge labor surpluses created by thousands of bracero workers, these perpetual cries for help sound comical in retrospect, but the system was not perfect for all industries, especially Spreckels's beet sugar interests in the Salinas Valley. Bracero labor was perfect for thinning beets, and beet growers came to rely on this source of labor. But as long as beet growers continued to use bracero labor for thinning, Spreckels's hopes of fully mechanized beet production went unfulfilled. If there were no braceros, though, valley growers might not be inclined to grow sugar beets at all, instead choosing to grow the more lucrative vegetable crops. Therefore, although bracero labor hindered the full mechanization of beet production, the beet sugar industry remained one of the Bracero Program's most faithful defenders, lobbying for its continuation every year. Spreckels found itself in the awkward position of promoting the advantages of mechanized production while still lobbying for the Bracero Program.

Cooperation between Spreckels, the UC, and machinery manufacturers provided solutions for the more technical problems of machine thinning. New technologies were developed that mitigated many of the problems involved with the new style of thinning. For instance, the UC worked with chemical companies to test new herbicides for controlling weeds. Also, agricultural engineers with Spreckels, the UC, and farm equipment manufacturers worked on new seeding technology for planting beets more precisely. This integrated approach, of course, was much more complicated than just having a crew of skilled farmworkers do the work, and articles on machine thinning in the *Bulletin* throughout the 1950s and early 1960s repeatedly emphasize the importance of careful planning for successful machine thinning.

Spreckels also attempted to portray hand labor as inefficient and scarce while at the same time promoting the rationality of a mechanized approach. For instance, *Bulletin* articles continually referred to farmworkers as scarce and of "uniformly poor quality," calling to mind growers' problems with high labor turnover and stereotypes of the inherent laziness of farmworkers (Spreckels 1965, 88). One 1951 issue of the *Bulletin* challenged California beet growers to replace bracero labor with machine thinning, and an accompanying article used a combination of pride and fear to prod growers in this direction:

California traditionally boasts about her leadership in mechanical harvest, yet she is at the bottom of the list in spring mechanization. . . . Why have California growers ignored these trends [toward machine thinning]? The answer, voiced in unison by most growers, would be that there was always an abundant labor supply at beet thinning time. . . . The traditional spring labor pool is now drying up. No longer can the beet grower count on abundant thinning and hoeing labor. With this warning, the beet grower can be expected to mechanize his spring work with the same determination and resourcefulness that marked his mechanization of the harvest. (Armer 1951, 11)

Another article, published three years later, used a "grower's own story of complete mechanical thinning" for further incentive:

"Why did you decide to use mechanical thinning on 41 acres, Rob, when you knew there might be some sacrifice in quality and when thinning labor is generally available?" I asked.

Rob answered, "You are wrong on both counts. Last January I saw that thinning labor might be hard to get—at least I didn't dare count on getting together as many men as I would need for my 150 acres. In the second place, now that I have thinned 41 acres by machine, I can't say that the results are the least bit worse than hand thinning."[33]

With its persistent crusade to fully mechanize beet production in California, Spreckels framed labor as a problematic part of sugar beet farming, which, as I have noted, was not a novel idea in California farm history. The discourses Spreckels used in this push, however, blended different ways of framing the labor problem that had not been combined before. Like the critics who challenged California growers' exploitative use of farm labor, Spreckels tried to portray dependence on hand labor as irrational and risky. By asking questions such as "Can you replace this

CAN YOU REPLACE THIS MAN?

The answer is yes — if you find labor scarce or want to cut thinning costs.

COLORADO
MICHIGAN
NORTH DAKOTA

and other states have mechanized the thinning of up to 47% of their

acreage — with substantial benefits to their growers.

Figure 4.8

Illustration from *Spreckels Sugar Beet Bulletin* challenging growers to replace bracero labor with mechanized thinning. From Spreckels (1951). Courtesy of Monterey County Department of Parks.

man?" (figure 4.8), Spreckels framed hand labor as an outmoded and dispensable practice of the past, denigrating the intellect of growers who still used hand thinning and harvesting. Unlike the critics who questioned the rationality of labor relations in California agriculture, Spreckels had to be careful to avoid portraying labor use as a social problem with unfortunate "social consequences." Instead, the company needed to balance its rhetoric and frame labor as a threat to rational farming practices without unduly offending growers or sabotaging its own efforts to continually renew the Bracero Program. By adopting mechanized production, Spreckels promised, beet growers could ally themselves with all right-thinking and industrious growers and thereby solve the farm labor problem.

Spreckels did finally achieve its goal of fully mechanized sugar beet production. By 1965 most growers in California and other beet-producing areas of the country were using machine thinning and harvesting. Not at all coincidentally, 1965 was also the year in which Public Law 78 was terminated, ending the Bracero Program. Growers howled in protest and predicted the death of all agriculture in California if bracero labor were taken from them, but most farm industries survived and in fact thrived, either by mechanizing production or making more use of domestic and undocumented sources of labor. The sugar beet industry made the transition to fully mechanized production with enormous help from the UC, which deployed new advances in seed breeding, seed planting, and weed control technology to reskill the practice of sugar beet farming.

In this respect, Spreckels did make a transformative change in beet farming labor practices, but the changes took place over almost 30 years, were tightly circumscribed by the larger politics surrounding farm labor, and depended on aid from every level of the state. Even if the use of Mexican nationals had been halted quickly at the end of World War II, it is doubtful that increased use of thinning and harvesting technologies would have truly solved Spreckels's production problems in the Salinas Valley. Figure 4.9 provides indirect evidence for this claim, showing a steady decline in sugar beet acreage after the Bracero Program ended in 1965, while the Valley's top vegetable crop, head lettuce, consistently advanced in the same period.[34] Today, there is virtually no sugar beet production in the valley, and the Spreckels Salinas factory, once the largest beet sugar processing facility in the world, closed in 1982 (Peterson 1981).

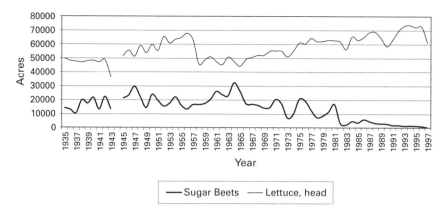

Figure 4.9
Acreage of head lettuce and sugar beets, Monterey County, California, 1935–1997.
From Annual Crop Reports, Monterey County Agricultural Commissioner's Office.

Mobilizing Science and Technology in a Time of Crisis

In some respects, the choices actors make between maintenance and transformation as modes of repair seem obvious. Actors, such as growers, with
a great deal invested in a complex system of production will make a
"rational choice" and attempt to limit the extent of repair to the minimum.
Those actors with less at stake in the system can more easily call for radical
change. But a closer examination of these decisions makes the nature of
the choices more complicated and less obvious. For example, what I have
described here as a program of maintenance for the growers—the Mexican
National and Bracero programs—were huge undertakings, involving international agreements, complex transportation issues, and worker training.
Similarly, the use of bracero labor, while certainly helping growers to
maintain and consolidate control over labor and further resist farmworker
unionization, did not eliminate conflict. As noted, Cooperative Extension
advisors in Monterey County had encouraged growers to raise wages during
the later years of the war as a means of stabilizing labor and preventing
walkouts. Why, then, were growers so uniformly insistent on maintaining
their use of migrant and bracero farm labor?

One part of the answer lies in the ecology of power of niche market
farming, especially the process by which power is created through production. Although the complex strategies employed by the growers and the

state were large-scale projects, in terms of the way growers produced crops, the creation of new organizations or the use of bracero labor changed very little. Growers still had large crews of workers thin their beet plantings and harvest their lettuce. In terms of repair and the production of power, the use of bracero labor during World War II and after allowed growers to maintain their production practices as closely as possible, thereby maintaining power over the production process. By tracing the practices and the lengths that growers went to maintain them, we can see how local practices and power are very closely tied together.

Further, the Spreckels case shows that the counterexample to this general course of action actually helps to support my analysis. Spreckels's competition with vegetable crops and its difficulties in securing enough beets to run its factory at full capacity pointed to a fundamental problem with the political economy of beet farming in the Salinas Valley. Because beet farming shared so many of the same production practices and costs with vegetable farming, Spreckels sought to revolutionize the production of beets through mechanization. However, by the time beet growers had fully mechanized production in the mid-1960s, the vegetable industry was very well established in the Salinas Valley and had already driven out many other extensive crops. In this respect, though the Spreckels factory in the valley operated for more than 80 years, Salinas was never really an ideal place for beet production and processing. In other beet-growing regions where alternative crops were not as lucrative, Spreckels had a much easier time convincing beet growers to make a transformative change in beet production. Therefore, Spreckels's attempts to transform the production of beet farming represents a special case of the relationship between practice and power, where a stable order never truly coalesced, despite much effort.

The production practices in which growers invest to raise crops, then, are vital for understanding how they define problems and plan for repair. Just as in natural talk between two actors, where conversation analysts have discovered that we have preferences for when and where we engage in an act of repair,[35] the structure of production (in the broadest sense) also gives actors preferences for how to maintain and protect this system. However, this does not mean that actors have a mechanistic or overly structured response to "external" problems. Instead, the nature of the problems themselves are contested and shaped through negotiation; in the case of farm labor, growers, farm advisors, and critics of California agriculture all found ways to identify labor as a kind of problem, albeit in different ways. These conflicting views forced the divergent camps to respond, as, for example, growers did when they created public relations campaigns to promote their definition against more critical voices. In sum, these negotiations represented a balance of structured practice and calls for change.

Despite their importance for understanding the story here, these seemingly mundane local production practices—and their role in the production of power—are only one part of the explanation. Growers were not alone in their efforts: the state intervened on behalf of growers at every level, from the local to the international. Without this assistance, it is unclear whether labor relations would have looked the same to the farm industry. As noted, those experts and policymakers in state institutions who worked closely with the farm industry during the war, including farm advisors, often expressed a kind of ambivalence about this relationship. Thus, the policies and practices of state agents also clearly reflect a kind of negotiated order, where the interests of the state, the farm industry, and other voices struggled to shape the form state intervention would take. In some cases, this negotiation caused tension among farm advisors and growers, as when advisors in Monterey County tried to reconcile their state-mandated goal of stabilizing the labor system and growers' attempts to keep wages as low as possible.

Overall, however, this conflict between the state and industry should not be overemphasized. Although in 1946 UC Director of Cooperative Extension B. H. Crocheron referred to advisors' activities during the war as a "rude interruption," the larger picture of the war years and after shows a very close and cooperative effort between farm advisors and growers. In this respect, farm advisors make an interesting case for understanding how state-based scientific and organizational resources are mobilized in a time of crisis. Cooperative Extension was a kind of organizational technology for growers, providing them with new practices and technology related to just about every aspect of labor, from labor contracting, to understanding the racial and ethnic "characteristics" of farm laborers, to the development of new machinery and other farm production technologies. Growers were successful in controlling their ecology of power by tapping into this technology, using it to shape the farm labor politics of California during war years but also the decades-long "crisis" described in this chapter.

5 Making a Place for Science: The Field Trial

Nature's Moving Target

Agriculture literally means "field cultivation," a blend of land and the practices used for growing crops in a particular locale. These two fundamental elements of agriculture—a local farming place and the work of farming—represent an ongoing balance between intervention and adaptation, where growers try to control nature to their own ends, namely, producing food. If modernity has changed anything about this relationship, it is only in the overall balance between adaptation and intervention. For example, although one grape grower acknowledged that there was nothing he could do about the unusually high amount of rain falling one year in the Salinas Valley and claimed that this helplessness "keeps them humble," most growers, with their predilection for research and intervention, seemed to agree with the attitude of another grower, who said, "Any time you're working with Mother Nature, you have a moving target." In order to grow niche market crops on an industrial scale, growers have invested a great deal of effort in this kind of intervention, transforming the Salinas Valley into a unique "place"; furthermore, this transformation is ongoing, as growers adjust to climatic factors and pest pressures that can change daily.

In many respects, a farm advisor's work represents how he or she would like to participate in this process of intervention. An extension program is a program for change, and advisors must also tailor their work to the unique land and farming practices in their assigned county. In addition, advisors' intervention has the added complication of the growers themselves: they must somehow convince growers to change *their* practices on *their* land. In this chapter I describe farm advisors' efforts to change agriculture in Monterey County through these two interrelated factors: the practice of farming and the place where it happens. More specifically, I focus on a form of experimental demonstration that advisors often use to simultaneously collect data on agricultural methods and change practices among their agricultural clientele: the field trial. Field trials retain the properties of other forms of experiment, such as control groups and special experimental methods, but are often conducted on a grower's property, and the data from the experiment is a grower's crop. Unlike laboratory-based experiments, which are intended to cut through the messiness and contingencies of place, field trials are intended to be place-bound.[1] In this respect, field trials combine agricultural and scientific modes of production: they produce crops as data in order to, in turn, produce consent. With this combination of features, advisors use field trials to "make a place for science," controlling a farming place through experimentation but also trying to maintain the particular "authentic" character of a given field. This authenticity makes field trials a powerful demonstration for growers, but the local, place-bound qualities of field trials also make them difficult to control; in many ways, advisors also need to strike a kind of balance when using field trials as a means of intervention.

   Because field trials combine these issues of place, control, and consent, they make a useful case for exploring the negotiation of order in agriculture. I focus on these aspects of field trials to show how they are a technique for repair. When advisors make a place for science, they reconstruct the relationships between people and things, altering the sociomaterial context of farming. By combining the kinds of work that scientists and farmers do, field trials also combine the knowledge that each gains through their work. This symbolic capital, in turn, allows advisors to argue for a new way of farming, thereby fulfilling their mission to change farming places, shaping how they are built. In addition, because field trials are ultimately grounded in the kinds of agricultural and scientific practices that growers and advisors use, field trials make an excellent case for exploring the relationship between practice and place. Field trials show how places are more than just a collection of soil, climate, bugs, and other local conditions; places are partly constituted by the practices that happen there. If science is not crafted to account for current farming practices, then growers are unlikely to accept the research-based advice of advisors. This terrain of practice and place, however, is sensitive ground. As I detailed in the previous chapter, growers' power is rooted in the interaction of labor, farming and technical practices, and the kind of capital that flows from this mix. Changes in practice can lead to changes in this relationship. Therefore, the prominence of practice in field trials gives them a dual character: powerful but also threatening.

This characteristic makes advisors' use of field trials a tricky kind of balancing act, where they must negotiate the boundaries of order and change. In this chapter I focus especially on this process and how advisors try to balance accommodation and control. Advisors can give their research trials an aura of realism and commercial relevance by placing them in a grower's field, but this also means special risks to the experiment's scientific status. This dual character of field trials leads to issues of control in the field. The diverse group of people who work in the field have varying interests, and their definition of control is different from that of advisors: namely, advisors want to control for a certain variable, and growers want to control anything that limits production. The struggle for advisors is to regulate growers' and other farmworkers' activities in the field for the sake of making a trial "good science" as well as commercially authentic. In sum, field trials bring together several elements from the ecology of power, providing a way of seeing in close detail the local negotiations between advisors and growers over farm practice.

Extension Work and the Importance of Place

The history of Cooperative Extension is closely tied to the history of field trials. When the idea and practice of extension work first began to develop in the 1890s and first decade of the 1900s, the concept of extension was nearly synonymous with the concept of demonstration, and extension workers were sometimes called demonstration agents. Seaman Knapp, who was instrumental in creating one of the first networks of extension agents in Texas during the early years of twentieth century, firmly believed that growers would be most receptive to new practices if they were demonstrated in the very place where growers farmed their crops (R. V. Scott 1970; Danbom 1979; 1995). Local demonstrations meant that growers could be more certain of the appropriateness of a new technique or technology for their farming place.

This connection between consent, place, and practice continues to this day, and the growers I talked to were very outspoken about the need for local research to address their problems. Often, they explicitly stated that scientists on the university campuses should not expect that research developed at the university would be accepted in the field. This grower chastised UC scientists and administrators for paying too much attention to campus-based, basic research:

Grower: If I only had one request it's that they understand that, in order for them to really perform the work that's gonna benefit the people of [this] valley, they need to do the work in [this] valley, not at . . . [the university]. They can do all the work they want up there and it may be handy for them [to do the work there]. But it doesn't solve things here. In my opinion, for the problems that are here, they'll have a hard time selling the results, if [the research] is done up there.

CRH: Selling it to people in the [farm] industry?

Grower: Right.

Other growers made similar complaints about the UC's research priorities. In part, these statements were likely attempts to exert control over the UC and influence the allocation of its resources. But growers' claims about the importance of place came up too often to dismiss as simple funding machinations. In the previous excerpt, the grower makes specific reference to his place, the valley where he farms, and he explicitly ties the importance of place to the ability to "sell" or convince growers of the utility of the research. What makes place so important for growers, and why are field trials useful for making new research convincing to them? In part, because they view their area as having unique soil, water, and climatic characteristics that influence the applicability of any innovation. In addition, growers understand that farming practices already in use are an integral part of farming success.

In this section I examine the importance of place in more detail, theorizing the connection between farming practices and farming places. To sell growers on a new way of farming, advisors must actively construct the preconditions for a new farming practice or technology to mesh with the current system of production. This is where field trials come in; field trials combine the control of experimentation with the unique particularities of a given place. This combination gives them epistemic authority with growers (sometimes) but also makes field trials hard to control.

As noted in chapter 1, place is a topic of increasing interest in several areas of social science that study the interface between people and nature, particularly science and technology studies (STS) and environmental history. Despite this growing literature, however, place remains difficult to theorize; given that everyplace is "a place," what kind of analytic power can place provide?[2] For my purposes here, the most fruitful way to theorize place is in relation to the practices that happen there. Practice is the linchpin in the larger ecology of power that structures farm work—it links the organic (land, plants, and bodies) to the institutional (commodities and knowledge) (see figure 1.2). Places are not simply aggregated practices, but practices, both agricultural and experimental, *account* for place; they are tailored to the particularities of the local. Conversely, places are shaped by the activities that people do there. Place and practice therefore have a dialectical relationship, and anyone who hopes to intervene there must become a good student of this local ecology.

As an illustration of this relationship between practice and place, consider the diagram in figure 5.1. The viticulture advisor showed this diagram to a group of growers while discussing his research on rootstocks for grape plants. Rootstocks form the base for grape plants, from which the vines grow and produce grapes. The viticulture advisor tested many different rootstock varieties in growers' fields across Monterey County in order give grape growers better information when choosing varieties for their own use. He was troubled, however, when growers asked for his opinion on the "best" rootstock variety. The advisor used the diagram in figure 5.1 to emphasize the many characteristics of a given field that influence the choice of rootstock. For example, some rootstocks have better natural resistance to phylloxera, a very destructive disease of grape rootstocks, and so these varieties may be a good choice for fields where phylloxera is prevalent. These disease-resistant varieties, though, are likely to also have downsides—perhaps poorer grape quality or lower yields. His diagram therefore stressed growers' need to balance competing factors when choosing the "best" rootstock for their vineyards. In the end, there is no best rootstock, and hard choices must be made to account for various contingencies of place and the practices that growers use for growing grapes.

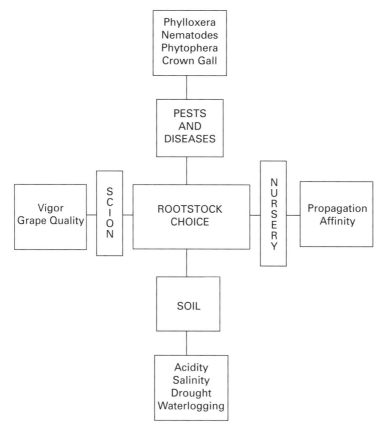

Figure 5.1
Diagram from a viticulture advisor's presentation to growers. Courtesy of UC Cooperative Extension, Monterey County.

This diagram, in conjunction with experimental data from the viticulture advisor on different rootstock varieties, is not meant to cut through the contingencies of place but to account for them, to make them the central factor in a decision. Of course, the diagram could potentially be even more complicated if it also had to represent the multitude of other factors that link the local ecology of a place to larger-scale aspects of its political economy, including the cost of different rootstocks, market conditions for certain types of wines, the availability of pest controls, and so on. Adding these considerations would greatly expand the diagram's size and complexity, but these are all additional considerations that determine the choice of the "best" rootstock. Although advisors are not given to mapping out all these factors, they constitute the larger ecology, are implicit in growers' decisions, and are issues that advisors need to confront when making a place for field science.

Without an appreciation for the practices already in place, it can be hard to understand why growers might not accept a seemingly superior way of farming. The advisors I worked with in Monterey County often commented on this link. In an interview with a soil and irrigation advisor, we talked about his efforts to reduce fertilizer use among local growers:

Soil/Water: Basically what we're looking at is going out to one of [the growers'] own fields and reducing the amount of fertilizer applied . . . in that field. Now, [if] they come back at harvest and there's no difference— you get the same yield, looks the same, storage life is the same—everything is the same, but you put half the fertilizer. They see that with their own eyes . . . that's valuable.

They saw it on *their* ranch, and that's the biggest barrier that you'll see: "Oh, that works on his ranch, but it won't work on my ranch." That's the story you always get. Because it is true—every ranch is different. Soil types are different. You know, the . . . irrigation system is different. . . . It would be nice if everybody had the same kind of soil and the same water and the same irrigation system. Then you could say, "Look, this works here." But you can't. It's not that easy.

Here, the place and the practices are inseparable—soil, water, irrigation method—and growers are unlikely to take much stock in advice that does not account for the interconnections between these factors.[3] Thus, field trials in growers' own fields are used to display the superiority of a new technique or technology in a setting that accounts for all the other practices that growers are already using. By appropriating both the supposed objectivity of science and the place-bound aura of farming a specific piece of land, advisors use these trials to build consensus around their research.

Typically, the form of field trials is fairly standardized. At the start, the farm advisor has an idea that he or she would like to test and demonstrate to growers. Then the advisor asks a grower for help, in the form of land, labor, and materials. The level of support from the grower varies from project to project, but the basic principle of the trial is usually the same: to incorporate the practices of growers and demonstrate the effect that just one change in this system of practice can have on farm production. In this respect, field-based experiments have much in common with laboratory-based forms of science that intend to demonstrate some phenomenon by controlling a variable or set of variables in a regulated environment. Figure 5.2 shows one such demonstration from the early days of extension work in California. Each photo portrays different parts of a field trial on potatoes, showing the benefits that fertilizer applications can have on yield. In this case, the part of the crop treated with fertilizer, in the bottom photograph, has a much higher yield than the unfertilized portion, in the top image. The marriage of science, place, and practice is complete here—the experiment changes just one practice (fertilizing) and holds all the other factors of this place constant. The tools of this field trial are not glamorous: old boxes and bags hold unremarkable potatoes. And yet, there is a powerful symbolism in this comparison. It signifies that something important and official has happened in this otherwise ordinary field of potatoes—the potatoes have become data.

At the same time, because it is in a seemingly typical field, the trial acquires a "pseudo-commercial" status, as the weed science advisor described it:

WeedSci: [Growers] loved to see something pseudo-commercial. These little small plots . . . they didn't have much confidence in those. You know because maybe [I] went out there and picked the weeds out to make sure it looked good or something [both laugh]. But once it was out there in the [grower's field], and they saw it standing up tall . . . compared to the standard, then they'd believe it.

The weed science advisor emphasizes two things that are also evident in the potato trial pictured in figure 5.2: field trials capitalize on the differences *between* two treatments, each *within* a grower's field. On one hand, this combination of the two key features of field trials—their emphasis on demonstrative comparison and their attention to the uniqueness of place—gives them epistemic power and commercial relevance. On the other hand, this very same combination makes trials tricky to pull off.

Figure 5.2
Results of a potato field trial, early 1930s. Potatoes from an unfertilized plot (*top*) and from a fertilized plot (*bottom*). The total yield from the fertilized plot is higher, with about half as many culls (discarded potatoes). Courtesy of Monterey County Department of Parks.

Making a Place for Science: Cooperation and Control, Part I

Field trials are designed to account for places, but their blend of experimentalism with place-specific features brings up important issues of *control*. Field trials are sites of heterogeneous practice and interest; the line between a controlled and uncontrolled setting is blurred in the field, creating a tension between the need for cooperation and the desire for control. Advisors and growers each want to control what happens in the field, but not always for the same reasons. Steven Shapin's study of the foundations of the laboratory and its image as a unique space for science makes an interesting contrast here. Shapin goes back to the seventeenth century, when Robert Boyle and other members of the Royal Society were establishing a new, experimental form of science. One unresolved aspect of the early experimental life was access to the setting for experiments—who could come in, and what behaviors were appropriate for this space? Eventually the lab was cordoned off from the rest of the world, both physically and symbolically (Shapin 1988b; Gieryn 1998). Experiments in the field, though, are difficult to sequester in this sense, both in terms of the people and the things that might enter the field and disturb a trial. Agriculture is a constantly changing intervention, and so advisors' work is partly shaped in reaction to this flux. Control over the field is negotiated and renegotiated with each instance of a trial and represents a considerable amount of worry and effort for advisors. In this section I describe how advisors cooperate with growers but also how each group attempts to control the other's activities.

To consider the problem of control in the field more fully, it is important to understand the relationship between advisors and the growers who participate in their research trials. In order for farm advisors to invest their research with an image of realistic, on-the-farm results, they need to gain access to growers' land. In addition to the realism provided by research in growers' own fields, there are other distinct advantages for the advisors from this arrangement. First, having the trial in a grower's field saves the advisors from having to grow their own crop as part of the experiment. Not only are the advisors and their staff nonexperts in growing a crop, but in many cases it can be particularly difficult to bring a crop to maturity. Thus, if the crop fails, it could cause the experiment to fail as well. Second, and related to the first point, grower practices tend to be relatively uniform across the industry and may be difficult to reproduce without appropriate equipment or technique. A third and perhaps most important reason is cost. A research trial can be expensive, especially if it involves destroying part of the crop. In the case of experimental treatments that have not been approved by the state for agricultural use, the treated crops must be destroyed immediately after the trial. In other cases, some part of the field trial's design may put part of the crop at a higher risk for damage from pests or other distress, sacrificing the marketability of the crop. Therefore, it is cheaper for advisors to put the burden of cost on the growers, although this limits the pool of growers who can cooperate in field trials to those who are financially secure.

For the most part, the advisors I worked with felt that they had relatively easy access to cooperation from growers on field trials. Advisors attributed much of this cooperative spirit to the "progressive" character of California growers (see chapter 3) and cited growers' appreciation for the benefits of research. But there are also some advantages that growers can gain from having the research performed on their own land. For instance, this ensures that research findings will be especially appropriate to the soil, drainage, and microclimatic conditions of their locale. In addition, growers are likely to welcome any improvements to their land that might come with participation in a field trial. For example, a trial testing new irrigation techniques may require that a grower's land be graded and leveled before the trial begins. Once the trial is over, the grower may choose not to use the new irrigation technique, but the land will still have been improved to perform the research. These potential benefits, which are not by any means universal in field trials, are weighed against the risks and costs associated with cooperation. Risks are created for both advisors and growers from the very combination of science and practical application that makes field trials so useful in the first place.

A key feature of field trials is a comparison between two different types of treatments in the same field. Typically, this is a variable or set of variables that the experimenter attempts to control, or hold constant, through the course of the experiment. Figure 5.3 shows a technician who worked for the Monterey County farm advisors marking off sections of a celery field with yellow tape that says "Do not pick" in English and Spanish. The areas being marked off represent an experimental crop area and a control crop area, the former treated with environmentally friendly chemicals and

Figure 5.3
A technician marks off sample plots from a celery field trial before harvest. Photo by author.

the latter with the conventional chemicals usually applied to this kind of celery crop. The idea was to compare the differences in yield and past control between the control area and the new treatment area. If the alternative chemicals proved effective at controlling pests and still allowed the same crop yield, this information may have convinced growers to use the new treatment.

There are several different meanings for *control* here, but two are especially relevant: the advisors typically *control variables* in their experiments, and the growers and advisors want to *control for pests* of their crops. The difficulty of maintaining a field trial on a grower's land lies in a potential conflict between these two kinds of control. Field trials often bring a level of risk to the grower's crop, depending on the research design. In some cases this may lead the grower to become nervous as he or she eyes the portion of the field in which the experimental treatment is being tested. For instance, in the celery trial depicted in figure 5.3, if by the middle of the trial the alternative treatment appears not to be effective, the grower might be tempted to spray a more conventional treatment on the experimental area of the field. In this event, the grower would be controlling for pests, but the advisor's distinction between experiment and control would be ruined. The entomology advisor described this dilemma:

CRH: Sounds like you have to be pretty up-front with [cooperators] about what you need and what you'll do.

Entomol: And they're usually pretty up-front with me too because . . . for example, when I wanted to test some parasite releases, I said, "I want to take this block of your field and I don't want you to use any insecticides on it." So they say, "OK, we'll do that . . . we're even willing to let the insects get bad in that section but if those insects start spreading through the rest of the field, we're gonna spray you out. And that's the ground rules. As long as things are going okay, that's fine. But if it gets bad, you know, we're out of here." So, and that's fine—as long as they tell me up-front what their expectations are and what's gonna happen, then I can live with that.

The thing that gets frustrating is, when you set something up with someone and you do all this work, and they call you up and say, "Well, sorry last night I had to go in and spray it by air and spray all your plots." [And I say,] "OK, well, thanks for calling but you just ruined six months of work" [both laugh].

In practice, this double risk—to the grower's crop and to the advisor's trial—leads to a need for surveillance in the field, yet a third meaning of *control*. In my interviews with farm advisors and fieldwork with their technicians, there was a constant emphasis on monitoring the ongoing progress of a particular field where a trial was being held in order to make sure that the grower was not doing something that would wreck an experiment. Advisors and their technicians make regular visits to the sites of field trials to examine the progress of the experiment and to monitor any dangers to the trial itself. This surveillance is particularly intense just before the crop is harvested because this is when data about yield and other crop factors are "reaped" by the researchers. Harvest scheduling is complicated by several factors, including prevailing market conditions for the crop, weather, and the quality of a given crop at maturity compared with others owned by the grower. Growers will often change harvest plans at the last minute to account for one of these factors. As the soil and irrigation advisor explained, this uncertainty makes harvest time a critical period for surveillance:

Soil/Water: You need to . . . make sure that the grower doesn't harvest [the trial] before you get out there. That's why it's important to have other people around—because, you know, you could call [the grower] five times and you go out there on the day before they're supposed to harvest and it's already been harvested. And that happens a lot. You have to realize [the growers have] got a different agenda than you do. And you're not generally on the top of their list as far as your experiments. . . . So you gotta really be on top of that all the time.

Only close surveillance of the field allows the advisors to control for these factors. At the same time, the growers themselves are keeping an eye on their fields and are wary of strangers who suddenly appear, tinkering among their crop. When I returned with some technicians to the celery field pictured in figure 5.3 for the trial's harvest, we were questioned by the grower about our presence in his field. This excerpt from my field notes describes the incident:

We . . . run into the grower as we park the trucks and get out. He is driving a pickup truck and has a dog with one of those plastic don't-lick-your-wounds hoods. He asks what we're doing here, and [Tech A] responds that we're from Cooperative Extension, taking some data from a field trial. [The grower] says something like, "Well, this is my field, and I don't know of any field trial happening on it." [Tech B], being the head of the research team without [the advisor] around, steps up to the truck and talks to him. [Tech B] says that we are [working] with x and y and z, spitting out the names of several people working on the project until one clicks. The grower says, "Oh fuck, I forgot about that! Good-bye!" and peels off in his truck. We all laugh a bit about that. [Tech A] says to me that the grower was probably worried that we were [county] celery inspectors.

This grower likely noticed the pickup trucks we arrived in, which have an official-looking seal on the door, much like trucks used by the county agricultural commissioner's office. The commissioner's office inspects crops throughout the county at harvest time and checks for potential regulatory violations or safety hazards. In addition to these inspectors, interlopers in the field can be taken for Immigration and Naturalization Service agents or even—in the case of pricey crops, such as artichokes or strawberries—for thieves. Consequently, advisors and their staff may have to renegotiate access to the field on each entry.

Making a Place for Science: Cooperation and Control, Part II

In addition to the advisors and the growers themselves, there are often other groups of workers who make their living in the field. This is especially true in California, where agriculture has long depended on migrant labor for planting, weeding, and harvesting crops (see chapter 4). The farmworkers working in harvest crews make for an even more complicated way of thinking about control because their skills are an essential, if somewhat obscured, part of an agricultural research trial in the field. Therefore, many of the practices that constitute a farming place in the Salinas Valley are actually the practices of farmworkers. Farmworkers help standardize the collection of data, making the data "scientific." At the same time, their skills make the trial seem relevant to the current standards and practices of the local farm industry. In this way, the work and skill of farmworkers are at the center of the production of knowledge for many field trials, just as they are for the production of commodities.

The part that farmworkers play in field trials mirrors other research on the role of the "invisible technician" within the literatures of science and technology studies and the sociology of work.[4] These works emphasize an irony about scientific practice: although technicians' work is indispensable for successful science, one rarely finds any evidence of this importance in official representations of research practice (published journal articles, etc.). The farm advisors' technicians suffer from this same invisibility, but their names do occasionally make their way into published accounts as co-authors with the advisors. The dualistic nature of this work—essential yet isolated—is further magnified in the case of farmworkers' participation in field trials. One does not often hear agricultural fieldwork described as skilled labor. Indeed, the work is perhaps the most denigrated and least envied of any occupation. But many of the niche market industry crops in California remain highly dependent on manual labor, including the vegetable and berry industries in the Salinas Valley.

Through this work, farmworkers develop a very deep knowledge of farming practices and criteria used for judging the maturity and quality of crops, and this is where they make a contribution to field trials. In order to avoid introducing extra variables into a field trial, the research design requires uniform farming practices; without them, it is impossible to obtain data that can be used for making comparisons between experimental and

Figure 5.4
A technician watches a farmworker cut and grade a plot of celery. Photo by
author.

control treatments. For instance, a common measure used for comparison
is the yield of the crops in the experimental and control groups. If the
harvest is not standardized, this may introduce errors into the yield data.
In many cases, these harvest practices are performed by farmworkers. Just
as the farm industry depends on farmworkers for a skilled harvest, farm
advisors call on farmworkers to help them get standardized data that are
relevant to commercial farming standards.

Farmworkers' skills and their importance for field trials is crucial yet
subtle. I use another episode from my fieldwork, involving the same celery
trial depicted in figure 5.3, to detail this contribution. Figure 5.4 shows the
harvest under way, and a technician (holding the clipboard) watching as
a farmworker cuts and grades heads of celery from one of the field trial's
experimental plots. The following excerpt from my field notes describes
the interaction between the advisor's research team and the farmworkers
that day:

[The research team] had to run to the [grower's store] quickly before we left so that
they [could buy] some "celery knives" for the harvest. These knives have a short

handle and a straight 8-inch blade with a crescent-shaped blade welded onto the end. The crescent end is used to cut the bottom of the head from its roots, and then the straight blade is used to trim and top the head. Although the research team anticipated having to do this work themselves on the research sections, they were a bit wary of it because they were not experienced with harvesting celery. They said that, although they *could* do it, it would be a lot slower, poorer quality, and they would be unsure of how to grade the heads. They hoped that they could get the [harvest] crew boss to assign one of his workers to help them with the harvest of the research sections.

So [TechA] (who is fluent in Spanish) and [TechB] went to talk to the crew foreman, and they persuaded him to send a [worker] to the research sections. . . . The worker came down and began harvesting the celery with his special knife—just like the ones recently purchased by the research team. He moved quickly and efficiently, cutting, trimming, and topping the heads and then laying them in rows that mirrored their pre-harvested position. When all the heads in the section were cut, he began to place them in boxes that said "Classy Celery" on the side . . . based on the size and weight of the head. The boxes were [marked for] "24s," "30s," or "36s," based on the number of heads that could be fitted into one box. Generally speaking, the smaller the number, the larger (and higher quality) the heads. This was the part of the harvest that seemed to intimidate the research crew the most, because they were unsure of how to make those distinctions. . . .

I drive back to the office and bump into [the advisor] in the parking lot. He says he doesn't know what's going on with the celery harvest, and I tell him that I just returned. I tell him that they got a [farmworker] to help them, and so it should go pretty quickly. He seems happy about that and says, "You know, that's the kind of thing we could do ourselves, but it would take us about seven or eight times longer than the people that really know how to do it."

Although the advisor emphasized the time that the farmworker saved for the research team, the more crucial factor here is the ability to make distinctions between different-sized heads of celery while harvesting. The farmworker made these choices quickly and very accurately.[5] If a technician were to harvest and grade the celery instead, there is a good chance he or she would mistake 24s for 30s or make other such errors. These errors would affect the yield count for a given plot by over- or underestimating the sizes of the heads.

Further, the farmworker also made on-the-spot decisions about how much to trim off each head he harvested. A major influence on these choices is any damage to the celery stalks from insects or disease; stalks that suffer from major insect damage are likely to be chopped off the head and left in the field. The amount trimmed off each head, of course, affects the final packing size of the head, and because trimming is directly related to yield, this is another area in which the technicians' lack of experience would affect the harvest data if they did this work. Since the trial was meant to contrast two different methods of insect control, the farmworker's knowledge of what parts to trim off the head was a precondition for accurate data.

In addition, by standardizing the collection of data, the farmworkers help the advisors get the *right kind* of data. The point of the trial is to judge what effect different treatments will have on the yield of the crop. Thus, it is important to know how many of each size head came from each test plot so that the advisor can correlate different treatments used on each plot with this information on yield. Without the skills for harvesting according to industry standards, the advisor would not have convincing data. In the celery industry, yield is generally measured in terms of the number of boxes that a given acre of land yields. The research team could circumvent their lack of harvest experience by not grading the heads but instead simply taking an aggregate weight of all the celery in each plot. In this case, though, the yield numbers would not be as meaningful to a grower who is used to thinking of yield in terms of boxes, not weight. Thus, the data, when presented to the grower, might not be as persuasive, even though they were taken from his own field.[6]

Advisors have a somewhat conflicted dependence on farmworkers. After the celery trial I talked with the entomology advisor and asked about the importance of farmworkers for the trial:

CRH: It seems to me . . . that there is a lot of interaction with the [farmworkers] who are doing the grading and the cutting. That seemed like it was pretty important.

Entomol: Sure, because, basically, we're relying on their expertise to tell us whether these plants are acceptable to the market or whether they would meet the quality standards for the grower that they're packing it for. And the guys who are doing the actual cutting are the experts in that area—much more than we are. And so, it was important for us to convey to them what we were trying to do. . . . Oftentimes they're very interested in what's happening and what the research is. And unfortunately we try to tell them as little as possible about that, at least until we're finished. Because we don't want them to be biased in doing the work. So it's really nice to have someone [who speaks Spanish] to be able to communicate very clearly to these people exactly what we need them to do and why it's important to help us out and what they can contribute to the project. And then after we're done, after they've finished, the likelihood of them getting involved with another one of these projects is relatively slim, so, if they're interested we try to explain to them what we've done and why we did it.

On one hand, advisors rely on farmworkers to provide practices that standardize the data and make them commercially relevant. On the other hand, they need to control the farmworkers so that the data collection is accurate. The entomology advisor did not speak Spanish and could not communicate with the farmworkers, most of whom were recent immigrants from Mexico. Therefore, he relied on one of his technicians to explain the researchers' needs and supervise the work (see figure 5.4).

In chapter 4, I argued that understanding the link between practice and the production of power was vital for understanding how farm labor relations played out in the twentieth century; similarly, labor practices are also important for understanding how knowledge is produced in field trials. Although these two cases are not perfectly comparable, in each the working body is a site for control. Advisors use farmworkers' skills to account for the relationship between practice and place, but they also closely monitor this work. In this way, field work is almost completely invisible in the production of applied agricultural knowledge, but this invisibility is not a given—it is only made invisible through more work.

Representing the Field: Controlling Field Trial Results

Control of field trials is not limited to work in the field. How the results from trials are represented and used to convince growers is also an important part of producing consent and promoting change through field trials. Why should a grower change farming practices based on the advice of a farm advisor? Several advisors emphasized the power of field trials to produce consent; recall the weed science advisor's statement that "once [a trial] was out there in the [grower's field], and they saw it standing up tall... compared to the standard, then they'd believe it." But this "seeing is believing" power of field trials is not always enough; though they establish a foundation for agreement through their basis in place and practice, they do not in themselves produce consent. The demonstration method is meant to be a transparent way of proving a new technique or technology, but convincing growers to change the way they farm is not simply a matter of showing them the extra potatoes.[7] This may frustrate the advisors, especially when a grower easily dismisses trial results with a wave of the hand and an aphorism: "One robin doesn't make it spring, and one . . . [small trial] doesn't mean that you're 100 percent convinced." Further, even if a grower does concede that a trial demonstrates an improved way of doing things, he or she still might not change, especially if the new way involves extra costs.

Thus, control is also a problem for how a place is *represented* through field trials, and in this section I examine how advisors attempt to accommodate and control the representational demands of growers. A trial may be well designed and executed to account for the contingencies of place, but growers also use representational practices that must be controlled. Just as doing a field trial on a grower's land increases the commercialism of the trial but makes control problematic, so it is with representations of the field. Advisors try to establish the same kind of control over depictions of the field yet also need to accommodate these data to growers' representational practices.

During my research with advisors and growers, we often talked about this topic—what was convincing or not about some type of representation—and I was struck by both the uniformity and the diversity of examples. The uniformity appeared through three factors that were often used to represent the field and field trials: numbers, visualization, and growers' orientation to these representations through their trust in advisors' work and advice. These are all ways of framing field trials to indicate the promise of experimental practices for commercial agriculture. For instance, the potato demonstration (see figure 5.2) is a powerful visual and numerical representation portraying the effects of fertilizer on potato yields. Although advisors and growers often invoked trust, numbers, and visualization to explain their personal interpretations, they attached very diverse readings and merit to specific instances of these factors. For any given field situation, the same detail might be used to support an interpretation in one instance and to devalue a claim in another. This grower, for example, used trust to explain his faith in a retired farm advisor:

Grower: What we say is, [Retired Advisor] did great research, he never screwed up. If he says it's good, it's good. Let's go do it.

Later in the same interview, though, he invoked the "seeing is believing" power of a trial that compared a new chemical fungicide with some older treatments. In this case, visualization trumped trust:

Grower: While we trust extension . . . there's nothing like seeing it with your own eyes to, to say, "This really does work."

CRH: It was pretty clear in [that trial] that it did?

Grower: Like turning a light on in a dark room.

In itself, this diversity is not surprising—the interpretive flexibility of representations is a founding principle of constructivist STS.[8] But representations that come from field trials and are intended to influence changed practices have complex interactions with the current practices and interests of growers, in addition to their personal relationships with and trust in the advisors. When used to represent field trials, numbers and visualization are *representations of place*; their worth is contingent on the multiple interests at play in the field. Therefore, just as advisors try to simultaneously control and cooperate with other users of the field, they face a tension between accommodating growers' representational demands and shaping growers' perceptions of a trial's results.

To illustrate this tension, I draw on a few examples of numerical and visual representations from field trials. I discussed the issue of numbers briefly at the end of the previous section. In the example of the celery trial, farmworkers' skills were essential for getting accurate data, but they were also vital for getting data relevant to growers. As noted, if growers in the celery industry think of yield in terms of boxes, then these are the data that will impact them the most. Beyond the measuring practices of a particular industry, though, there are also the standards for individual growers and their companies. For example, when I asked one grower about the use of numbers in farm advisors' presentations at meetings, she described how different growers may have different numerical production standards:

CRH: You mentioned before that a lot of times the [advisors'] presentations are quantitative. That's something you think convinces a lot of people, too—they can see charts and stuff that have a lot of data in them?

Grower: Yes.

CRH: Because that makes it more clear or . . . ? Makes it more systematic?

Grower: Well, they can compare the numbers to their own standard. . . . Take, for instance, one . . . [advisor's] trial. [He measured] the number of [insects] per [plant] in all the different treatments. The grower can look up there and go, "Well, to me, five per [plant] is what I'm shooting for—that's acceptable to me, so that's gonna be the best treatment for me." Whereas another grower might go, "God, I don't know. I haven't heard any problems from our harvesting guys and I know I'm using [treatment X], so twenty per [plant] must be acceptable." And they're able to then gauge, on their own standard, how their practice is appropriate.

Each grower in this example had different standards for the level of pest pressure on a crop, but the advisor's numerical representations met the practical demands of both. By understanding the specific computational practices of growers, the advisors can make their data more persuasive (Lave 1988).

This effort at accommodation is balanced with a need to control the kind of representations gathered from a trial and how they are interpreted. During the period of my research, the entomology advisor was trying to get more growers involved in the design stage of field trials. He hoped that this would make growers feel included and spur them to adopt new practices more quickly; the celery trial (see figures 5.3 and 5.4) was part of this effort. This situation made the tension between accommodation and control very clear. The entomology advisor told me about the conflicting demands that he and growers have for representations of the field:

Entomol: What's kind of hard for me to realize is that [the growers] are talking about changing the pest control practices in the field and I'm interested in what that does to the insects. But, the bottom line is, they don't care about the insects. What they care about is yield. And insects do affect yield so ultimately they are interested in the insect population. But all they really care about is yield, and if it were left up to them totally, all they would measure is yield. But I, as a scientist, am in there trying to convince them that they also have to at least look at why that yield changed from one side to the other. Because if we don't know *that*, then we won't be able to say, "This is another situation where it's likely to work," or not likely to work.

The growers have a strong interest in place-bound representations, but the advisor finds these too circumscribed by their particularity to stand alone. Here, the advisor explicitly identifies himself as a scientist, arguing for representations that will allow him to make judgments about causality and applicability to other sites. For instance, the advisor would like to know how many insects are in the field near each treatment and whether this is correlated with yield. Other pertinent information can include data on the population of beneficial parasitic insects in each treatment. Once again, the balance between the scientific and commercial aspects of field trials is tricky to maintain.

Perhaps the most complicating factor about representations is the way they get mixed together in field trials. Trust, numbers, and visualization are part of just about every trial, but they may say very different things about whether something works or not. This is especially true in trials testing new practices that are meant to be more environmentally friendly than current techniques. For many years, field trials that tested chemical interventions in agriculture were simply focused on whether the chemical controlled a pest without damaging the crop. Of course, now advisors and growers (as well as the general public and government regulators) are more aware of the effects that agricultural chemicals have on the wider environment, and field trials are being used to test ways of farming that balance profitability and environmental effects. These field trials are meant to represent more than just whether "it works," and therefore their representations can be harder to control. This advisor describes his difficulty making growers "see" the utility of a more environmentally benign chemical:

Entomol: There's a botanical insecticide and it's called Neemex—it comes from the neem tree. It's an interesting product because it's very safe to humans and mammals, has very little impact in the environment and it acts . . . as an insect growth regulator. So, if you spray it on these insects they're not likely to die. But they're likely to never complete their life cycle and [reproduce]. . . . And that's something that's almost impossible for a grower to see just by casual observation and walking their fields. But I've been presenting data for about four years now showing that, you know, this happens, this really is something that you should consider using. . . .

   The approach has always been, if you see a bug, spray it and kill it. Well, that doesn't work very well anymore because the sprays we have don't kill the bugs, at least not in the old form. We've got to look more at managing [insect] populations, and that starts talking about area wide kind of management. We can no longer talk about just protecting this field. We've gotta talk about protecting the crops on this ranch. Or we talk about protecting the crops in this part of the valley. So that brings up interesting issues because you have growers who are treating their field, and they're saying, "Well, I don't see the impact here." Well, no, but in your fields that are planted around that, you'll have less insects.

   Here the advisor is working against a common way of representing the success of a field trial on insect control: looking for dead bugs in the test plot. The chemical he was using does not kill the insects but instead regulates their growth so that they will not fully mature and reproduce. Further, success in this trial is hard to represent through the traditional conventions of field trials; the "seeing is believing" power of field trials actually works against this application in two ways. First, the chemical is intended to work over a longer period of time, controlling insects throughout their life span. Admittedly, the insect that the advisor describes here has a short life span measured in weeks, but the impact of the treatment may still take several insect generations before the results are apparent. By the typical standards of a field trial, this is a relatively long period of time to wait for results, especially given that a crop's life span is also not very long. Even if, over the long term, numbers show that the alternative treatment did not hurt or even helped yield, the growers still see insects alive in their fields in the short term. Second, this chemical is meant to manage an insect population over a wider area; the advisor said, "We can no longer talk about just protecting this field." Again, this strategy is difficult to represent. The chemical may in fact be controlling the insect population on a larger scale, although specific fields may still have significant numbers of insects. Like the temporal issues involved in this trial, these spatial differences could be quantified by taking a kind of census of the area insect population, and perhaps such a survey would show lower numbers in the overall population. These numerical representations, however, still contradict the growers' demands for "seeing" results in accordance with the standard conventions and results of field trials. This interplay of different representations of the field makes environmental problems more difficult to address and represent through field trials.

As a final instance of the intermingling of trust, numbers, and visualization in field trials, I examine a case that was very much about "seeing" the results of the field trial. This case did have a kind of census of pest pressure through numbers and also brought in the issue of trust. In all, this final example depicts very clearly the negotiations that take place over land and its representation through field trials.

The case concerned attempts to control a disease called lettuce drop, caused by the *Sclerotinia* fungus. Affected plants droop and lie flat on the ground instead of forming round heads of lettuce. Lettuce plants infected with *Sclerotinia* are generally unmarketable, and therefore the disease can cause considerable losses. In the early 1980s an advisor (now retired) who specialized in agricultural machinery and engineering began testing a technique called deep plowing for controlling *Sclerotinia* on lettuce. Deep plowing requires a special plow that, at the time, was imported from Europe and was quite expensive to purchase or rent. Deep plowing works by turning the soil upside down, replacing the topsoil with 20 or so inches of the soil below it. This turning is intended to take the upper levels of the soil, where the disease is more prevalent, and replace it with the lower levels, where there is less sustenance for the *Sclerotinia* pathogen. The deep plowing leaves giant clods of soil in the field that have to be broken up, and growers have to replow the field several extra times. Based on the novelty of the idea, the advice and research of the advisor, and widespread damage from the disease, many growers in the valley began using this technique in an attempt to control lettuce drop.

About ten years later, in the early 1990s, the UC assigned an extension specialist in plant pathology to work at a USDA research facility in Salinas. During my fieldwork in Salinas, I spoke to this specialist, and he described one of his earliest research interests when he arrived in the area, the use of deep plowing to control *Sclerotinia* on lettuce. After ten years of using the special plow, he said, growers were still seeing mixed results and were puzzled. The principle seemed sound, but they were not getting the kind of control they had expected. Further, because the plow is expensive and the technique involves a lot of time spent on plowing and replowing a field, growers were investing a lot of money in this method. The specialist felt that growers were hungry for an explanation; they wanted help to make deep plowing work or at least a justification for why it would not work. In cooperation with the plant pathology advisor in Monterey County, the specialist was able to find a field where a grower was about to use the deep plowing technique. They took soil samples before and after the plowing to check for *Sclerotinia* and found that deep plowing had made the disease problem worse. Before plowing, *Sclerotinia* pathogens were generally limited to just one area of the field, but when growers replowed the field, they spread the disease that had been limited to one part into all parts of the field. After this research trial, the specialist recommended that growers not use the deep plowing method.

With this background in place, I turn to an interview with a grower in which we talked about the deep plowing technique. I include a long excerpt from the interview because it speaks to the interconnections of several different forms of representation and their situational use. In the excerpt, the retired advisor who first brought deep plowing to Salinas is referred to as Ag Engineer, and the extension specialist stationed at the USDA facility is called Specialist:

CRH: Can you think of any times when you or someone from your company took information from [an advisor's meeting] and applied it in your company?

Grower: Sure. We used to have [Ag Engineer] who was [in Cooperative Extension] and he was pretty well-known and his forte was equipment. He was doing some plowing work on lettuce drop or *Sclerotinia* and we bought some plows and started doing some deep plowing to kind of help this problem. Some of the newer work says that you shouldn't be plowing [for this problem], but I still feel that . . . it works. Like I say, there's a new scientist working out there and he feels that plowing just spreads the problem out on a broader base of the field. So—

CRH: Is that [Specialist]?

Grower: Yes, [Specialist].

CRH: But based on your experience it seems that plowing works.

Grower: Right. We still continue to plow and we haven't had a problem. I disagree with [Specialist]. And I disagree with him because I see what it does. He's doing it based on his method of grading the field and counting sclerotia like that.

CRH: But based on the lettuce that you see . . .

Grower: Based on the results that we've found I feel that [plowing helps].

CRH: And that was something that came from a presentation.

Grower: It came from [Ag Engineer]'s work. He was funded by the Lettuce Board.

CRH: So, what is it that you found convincing about his presentation that you don't necessarily find convincing about [Specialist]'s presentation?

Grower: Well basically we tried it and it worked. [Specialist] is just saying that it doesn't work. He's doing it through his trial work and, well it just hasn't caught on—[Specialist]'s philosophy hasn't caught on. . . .

He could be 100 percent right but, you know, we've had situations where we *haven't* plowed the fields for whatever reason—probably because we didn't have time to plow—and the following crop was certainly not better because we didn't plow. Plowing is not a real favorite tillage of a lot of people, because they feel it destroys the soil. . . . [But] it's a tool that's been around for years.

There are many different kinds of representations in play here. First, there is the visual: seeing how the plow works, being convinced by the idea of turning over the field, and seeing that "it works" when they plow and does not when they do not plow. Second, numerical representation is at work as well: the specialist has counted the number of sclerotia in the field before and after the plowing, but the grower discounts this approach. Third, trust is apparent: the grower discounts the specialist's advice as "philosophy" and contrasts it with his experience with the plowing technique. With the grower's trust already committed to the prior work of the retired advisor and the "idea" of the plow, the specialist's newer approach cannot get a foot in the door.

This example is not intended to show the power of one kind of representation over another but rather to illustrate how grower practices structure their use of these representations, and how they use them to make decisions about work on their land. Representations are played against each other and valued according to the current practices and requirements of farming in a unique place. Further, these can all shift from season to season, as changes in climatic, pest, and market conditions modify farming needs. Overall, these examples point to the surprising murkiness of field trials—a type of experiment that ostensibly produces straightforward, unambiguous results. When practices are put into their context, we can see more clearly why growers do or do not make changes based on the results of field trials.

Field Trials and Their Potential for Repair in Agriculture

Field trials are an interesting blend of order and change in farming practices. They incorporate a grower's current farming methods but also make small changes in them. Social order and change seem like opposites, but in this chapter I have argued that advisors must create a local infrastructure of order to successfully promote change. Of course, change can happen quite easily when things fall apart and people are free to do as they wish—anarchy makes change easy but hard to control. This brings up yet another aspect of control inherent in field trials: field trials are attempts at a kind of *controlled* change, confronting an established system of order and arguing for a relatively small change. Thus, the use of field trials to intervene in agriculture is a fundamentally conservative approach to change. By focusing on this type of change, field trials attempt a kind of repair as maintenance that may preclude larger, more transformative changes in agriculture. Recall the entomology advisor's field trial to demonstrate the efficacy of an insecticide that controls insects by regulating their growth cycle. This advisor wanted growers to adopt a new chemical, but he also hoped that growers would embrace a new way of thinking about insect control—one that considered the whole valley as the place for pest management. Growers were hesitant to accept this method of control because it did not fit with their current way of "seeing" a field and its insects. Getting growers to see in this new way will take a transformation, not maintenance, and field trials may not be as effective for this kind of change.[9]

In addition to taking a conservative approach, field trials conducted on growers' land may serve to reproduce inequality among growers, especially along the axis of social class. I occasionally heard advisors and others in the Salinas Valley farm industry claim that there "are no small lettuce [or broccoli, or grape, or . . .] growers here." But smaller farms do exist in Monterey County and especially in adjacent Santa Cruz and San Benito counties, where many of the advisors had cross-county responsibilities. Perhaps the best example of the continued coexistence of industrial-scale production and small farms is the strawberry industry. The marketing and distribution of strawberries on California's Central Coast is dominated by a few very large corporations, but a significant portion of the actual production of strawberries is done on small farms, often by former farmworkers who were able to buy or lease a small plot of land after several years of fieldwork (usually less than 50 acres). In some cases, it is the corporations themselves who lease the land to these small growers, in an arrangement that is essentially a form of sharecropping (Wells 1996).

Just as the rich and the poor tend to live in different places in urban areas, wealth and the quality of land are also closely correlated in the case of agriculture: rich farmers have good land, and poor farmers tend to work on land with inferior soils, water, and drainage. In this way, a field trial that tests a new technique or technology on the land of wealthier growers may not be as applicable to land farmed by poorer growers. As I noted in chapter 3, research conducted by Daniel Mountjoy (1996) shows that Mexican-American strawberry growers on California's Central Coast are less likely to cite Cooperative Extension as a trusted source of information when compared with white and Asian-American growers. Mountjoy cites cultural differences as the main barrier that prevents Mexican-American growers from taking better advantage of Cooperative Extension and other state-based resources. But if place matters for growers, and they assess new practices against an ecology that includes an expanded, sociomaterial conception of place, then research developed for one kind of place may not be as appropriate for another.

This is where the funding of field trials becomes an important factor in understanding the connection between place, trust, and change. As employees of the university and of the state, advisors' research results are meant to be a public resource, free and accessible to all. In this respect, field trials, as experiments meant to elicit consent through public demonstration, should be open to all, regardless of whose land and money are used to conduct them. Advisors put a great deal of effort into disseminating these results as widely as possible, through field days, meetings, publications, and other methods of publicity. In the end, however, the risk and cost of field trials on a grower's own land is a limiting factor in what kinds of places advisors can choose when conducting a trial, and these methods of dissemination will always be removed from the actual practice of the trial and its direct connection to place. In this way, advisors' field trials make a kind of place that is often most appropriate for the needs of the largest industrial growers. If nature is a moving target for intervention, then field trials are ultimately a form of maintenance that reproduces power, helping niche market growers adapt to their constantly shifting ecology.

6 Making Change: Power, Politics, and Environmental Problems

The Paradox of Repairing Environmental Problems

Agriculture is an environmental problem. This is a straightforward claim, but with many possible and interrelated interpretations. The first, and broadest, way of interpreting this statement returns to my conception of agriculture as a system of interdependent sociomaterial elements, where growers, farmworkers, scientists, and others work to manage the character and control the outcome of this interaction. In this way, the environmental problem at the heart of agriculture is inherent in the very idea itself—there is no way to avoid the fact that farming shapes and changes its local ecology in order to master it.

In contemporary usage, applying the term *environmental problem* to agriculture often carries a different but related meaning: a problem of consequence, where farming leads to negative impacts on the health and quality of its environs. At the start of the twenty-first century, growers and agricultural scientists face a paradox, a dilemma arising from these two ways of framing environmental problems in agriculture: the very practices and techniques that have allowed modern agriculture to take control of farming environments (the first set of environmental problems) have also created a range of destructive, largely unintended effects (the second). As just one example, the development and widespread use of pesticides and synthetic fertilizers was one of the most important agricultural innovations of the early twentieth century, allowing ever fewer growers to produce ever more food. At the same time, however, these technologies have been implicated in water pollution on a nationwide, even global, scale.[1] Growers and agricultural scientists now find themselves in a strange position: how can they change a system that they themselves constructed and promoted, maintaining the system while mitigating its environmental impact?

For farm advisors, addressing environmental problems is a kind of double repair—a repair of a repair—where they attempt to modify or curtail farming methods that Cooperative Extension promoted in the first place. It is only slightly an exaggeration to say that the ecology of power in niche market industrial agriculture was founded on the very technologies that are now most implicated in the environmental impacts from industrial-scale farming. Salinas Valley growers are largely dependent on very intensive use of synthetic pesticides and fertilizers, raising the question of whether and how much their use could be reduced. Whereas in the past a new pest could be addressed through the use of chemical interventions, environmentally aware solutions may require increased costs and more work on the part of growers. Further, the often controversial nature of environmental issues only exacerbates this difficulty. Growers in Monterey County and California are often portrayed as big fans of research-based intervention, but what happens when science says that agriculture is a threat to the environment? Conflict over environmental problems and agricultural research may then spill over into growers' perception of science and its roles in repairing agriculture. In these situations, just as the agricultural industry mobilizes to solve the kinds of problems described in the previous chapters, growers may deploy their political, economic, and organizational resources to attack the definition of agriculture as an environmental problem.

These tensions between established farming practices and discourses, impacts on the environment, and attempts at change are the subject of this chapter. Because environmental problems invoke both the practical structure and discursive meanings that compose agricultural production, they make an excellent case for exploring how repair encompasses these multiple levels of social order. Because many environmental issues are less obvious than the sudden damage from a swarm of insects or an outbreak of disease, there is often conflict over whether a problem exists at all, much less what the best solution is. This ambiguity creates an opening for actors to promote their own definition of a problem through rhetoric and influence, thus forming yet a third way of thinking about environmental problems in agriculture, as political problems. When growers and others resist the designation of agriculture as an environmental problem, claiming that such judgments are based on "politics, not science," they are working to shape the discursive context of what counts as an environmental problem for agriculture.

Whether for the sake of public relations, forestalling potential regulation from the state, or genuine disagreement, this way of repairing environmental problems may conflict with advisors' attempts to promote change through new practices and technologies. As with field trials and environmental change (see chapter 5), this conflict places advisors once again on awkward ground. How can they work within a structure and at the same time seek to change it? Given the political complications surrounding environmental problems, we would expect it to be difficult, and recent social theory on the relationship between scientists and industrial interests support this expectation. For example, Ulrich Beck's risk society thesis holds a very dim view of the ability of scientists to address environmental problems.[2] Beck claims that science and technology are themselves institutions at the heart of industrial modernity, responsible for generating environmental risks in the first place. Further, science and technology are often deployed as a kind of "'counter-science' gradually becoming institutionalized in industry" and beholden to the interests of capital.[3] In contrast to Beck's risk society thesis, ecological modernization theory places science and technology at the center of attempts to create a "green" modernity, presuming that scientists and other experts will, in concert with vast changes to the regulatory state, be a vanguard of environmental change.[4]

And yet, despite these divergent perspectives on the ability of science and technology to address environmental problems, there has been very little empirical research on the role of experts in environmental conflicts. Overall, science has been treated as an institutional black box in the theoretical debates over modernity and the environment. Instead of critically examining the place of science in specific environmental conflicts, many authors have assumed an overly simplistic view of scientific practice (Wynne 1996). Using the case of farm advising, my aim in this chapter is to demonstrate that advisors *do* act as agents of social change but also to show that their work in this regard is tightly circumscribed by the larger political economy of the farm industry and the regulatory state. As with the forms of repair I described in previous chapters, advisors are constrained by the institutional context of their work, yet they use this context strategically, manipulating elements in the ecology of agriculture to frame the perception of environmental problems and possible solutions. Although this strategy is a fundamentally conservative technique of social change, it also allows advisors to develop close relationships with some of the largest contributors to farm-based pollution.

In this chapter I draw primarily from contemporary interview data collected during my time in Salinas. I also draw on field notes taken during meetings that concerned environmental issues and corresponding research. The discussions I had with advisors and growers about agriculture and the environment were sometimes general but often revolved around specific cases of environmental problems in the valley and throughout Monterey County. In the next two sections I discuss the political and regulatory context of environmental problems in agriculture. I then move on to two case studies that explore these issues in more detail.

Defining Environmental Problems in Agriculture

If working on environmental problems related to agriculture is difficult, perhaps the simplest strategy for advisors would be to ignore them and instead focus on the kind of production problems that the industry has always called upon them to solve. This is actually the strategy that the previous generation of (now retired) advisors employed when attention was first widely brought to the environmental impacts of agriculture in the 1960s and 1970s, especially following the negative consequences of pesticides like DDT. Although the UC slowly began to respond to these concerns, it was not until the late 1970s and 1980s that the growth of the environmental regulatory state forced growers and agricultural scientists to take stock of their growing practices in the light of environmental impacts. In the same way that the UC and its farm advisors dragged their heels when responding to accusations of bias from racial and ethnic minorities during this period (see chapter 3), the prior generation of farm advisors could largely, in the words of one retired advisor, "pay some lip service" to environmental issues.

This is not a strategy that the contemporary advisors use. In fact, a great deal of the Salinas Valley advisors' time and resources is devoted to addressing problems related to the environmental impact of farming. In part, this change may be attributable to cultural change in science—the overturning of old paradigms, the rewriting of textbooks, the ascension of a new generation of researchers—described in Thomas Kuhn's (1970) classic account of scientific change. But a great deal of the attention that advisors give to environmental issues has to do with the issues' prominence as industry problems in their own right. The regulatory state, from county-level agencies to the federal government, is increasingly regulating agriculture. Each new discovery about the environmental consequences of agriculture brings the prospect of new regulations; like it or not, growers often face new restrictions on the techniques and technologies they use for farming.

For example, the U.S. Environmental Protection Agency (EPA) has identified agriculture as the largest single nonpoint source of water pollution in the United States.[5] Agriculture in Monterey County is no exception; at the time of my research, it was widely believed to be responsible for a high level of nitrate contamination in groundwater. Drinking water in wells throughout the county showed higher-than-healthy amounts of nitrate, most likely from excess synthetic fertilizers leaching out of a crop's root zone (the depth of the soil that a plant's roots reach to take up water and nutrients). High levels of nitrate, when consumed in drinking water, can interfere with respiration, especially in the very young and very old. Wells throughout the valley were designated unfit for drinking water, and the nitrate contamination issue received a lot of attention in the local press. Residents in some of the poorest areas of the valley were generally affected most by the contamination, and the county trucked in drinking water for their use. Local community groups petitioned the county to drill new, deeper wells, and the issue was eventually publicized in the *San Francisco Chronicle*, to the dismay of many in the farm industry (McCabe 1998a; 1998b). Because of the severity of the problem and the publicity, some kind of regulation seemed imminent, and the vegetable industry appeared to be one of the most likely sources of the contamination.[6]

In addition to the nitrate problem, the use of pesticides remains controversial. Ever since the publication of Rachel Carson's *Silent Spring* (1962), pesticides have been synonymous with environmental threats. Although, when compared to the nitrate problem, the use of pesticides in Monterey County was not tied as directly to specific environmental risks, there were significant pressures on growers to reduce their dependence on pesticides. In 1996 the Food Quality Protection Act (FQPA) was passed into law by the U.S. Congress. The FQPA was originally lauded by agricultural industry lobbyists as a boon for agriculture, mostly because the Act provided expedited approval processes for the use of new chemical inputs. These changes could potentially be beneficial for growers of vegetables and other "minor crops."[7]

The farm industry's initial feelings of goodwill toward the FQPA wore off, however, as the EPA began implementing another aspect of the FQPA's mandate: reviewing all chemicals registered for agricultural use and rethinking their acceptability under new guidelines. These new guidelines required the EPA to reexamine the potential for harmful aggregate exposure to chemicals with similar mechanisms of toxicity; potentially whole classes of chemicals could be deregistered based on the assessment of the EPA and an international committee of experts. If certain classes of agricultural chemicals were deregistered, the range of chemical tools available to growers of minor crops would shrink, leaving fewer or no controls for some crops.

I attended a meeting organized by a growers' association to educate growers about threats to Salinas Valley agriculture from the FQPA. The following excerpt from my field notes describes the apprehension with which growers greeted these events:

Another thing that really got them down was the "lumping" of [chemicals] into one broad classification. . . . [Speaker 1 read] a quote from the international committee [that is reexamining the chemicals, which said] there is no reason to have subcategories for a type of chemical called an organophosphate [OP]. At this, a woman sitting next to me sighed, and other people around the room seemed to fidget a bit. [Speaker 1] then said that they were further considering putting all the carbamates in with the OPs, and at this the woman muttered, "Oh shit." [Speaker 1] gave an example of [the committee's] logic: "The way they think about it is this: a guy is walking around his yard spraying [for weeds] and has his kid with him. While he's not watching, the kid picks up a big piece of sod and puts it in his mouth. Then the kid goes to daycare where they've recently had to spray for some insects, and the kid gets something on his finger and is exposed again. Then, the dad sets up an insect fogger in the basement for roaches and the kid manages to get down there and is exposed again." [Speaker 2] then adds, "Plus, anything that is supposedly in the food the kid eats."

If whole classes of chemicals such as organophosphates and carbamates were deregistered, many of the pest control tools used in Salinas Valley agriculture would be eliminated. As the drama and tension in this example indicate, the farm industry perceived this to be among their most pressing environmental problems.

These examples show the political context of environmental issues in Salinas Valley agriculture and highlight the pressures that growers perceived from regulation. In large part, growers were not worried about environmental problems as production problems; the developments I described would not have much effect on the yields they were getting from their fields. But they saw the threat of increased governmental regulation, the loss of certain chemical controls, and the negative publicity as a real menace. In this respect, it is not the environment that is a problem for growers but rather environmentalism. Therefore, farm industry responses to environmental problems are often just as much concerned with spinning a positive image of agriculture and deflecting potential regulation as they are about actually addressing the problems themselves.

This does not mean that growers are always indifferent to environmental issues. It merely reflects their struggles to maintain control over production practices and the larger sociomaterial ecology of their industry. Recognizing this is essential for understanding the tricky relationship between agricultural scientists and the farm industry with respect to environmental issues. Many of the growers I interviewed explicitly categorized environmental problems as political problems and cited the state as the source of unwanted—in their view, often unneeded—intervention in the practices of the farm industry. For example, this grower described testifying before a California state commission to seek approval to use a pesticide in his industry:

CRH: Were you involved with any other problems that weren't necessarily related to production—say, political issues or things like that?

Grower: Well, a little bit. When I first started in the pesticide field, we used to go to Sacramento and testify on why we needed [to use a pesticide]. . . . After doing this two or three times, I happened to have an opportunity to speak to one of the people sitting up in front of us—I guess you could call it the Board or whoever was evaluating [our request]. And I happened to bump into him after that, and I said, "What happened up there? You sat on the panel up there, and I'm just curious as to what happened. Because the state never did ask us anything, so what are we doing here?" [Laughs.] And this guy was a qualified person. He said something like, "Hey, they had their minds made up before you got here." Now that's [just] politics, [but] I didn't do it [again] after that.

CRH: It just seemed like a waste of time?

Grower: Well, to me it was a waste of time because . . . [the state] has their own people and they make their own decisions, but, according to the law, they have to have hearings. And I think they have hearings sometimes just [to satisfy] the law.

In another interview, a different grower implicated the UC system itself as being overly involved with the politics of environmental change. He complained that urban politicians with a dilettantish interest in nonconventional farming had better access to the UC than the farm industry did:

Grower: [The university will] listen to some dingbat from San Francisco who wants to have sustainable organic agriculture on the roofs of the apartment buildings in San Francisco. Because [the dingbat] will vote against you if he doesn't get that in. And so you will devote political time and devote extension time to do that, because that's where the vote is. I understand that's the site of the problem. You know, I just think that somebody in the hallowed halls of Berkeley or Davis could say, "Screw you—we're not doing that—go away." But they never do.

In each of these cases, the growers portrayed state actors as political agents who based their decisions on political convenience rather than on an understanding of the ecology of the farm industry. These complaints often made a distinction between actors inside the farm industry, who had knowledge of the local conditions of farm production, and outside actors, who did not. Typically, growers portrayed government regulators as well-intentioned but naive outsiders at best and as manipulative zealots at worst. These outsiders, the insiders argued, lacked a full understanding of the local conditions of agricultural production and wanted to change the ecology of agriculture without understanding how politics related to practice. Thus, the state's regulatory solutions were based not on information or familiarity but on uninformed, politically motivated decisions.

This way of framing environmental problems represents a problem in its own right for the researchers, especially the farm advisors, who are working on solutions to environmental problems. As employees of UC Cooperative Extension, the advisors are charged with improving certain areas of agricultural production, and mitigating environmental problems is a significant part of their efforts. At the same time, they are not an official part of the regulatory state, even though they are employed by the state. Growers' attitudes further complicate this situation: although advisors may feel that they are acting in the interest of the farm industry by working on environmental problems, there is a danger that growers might conflate advisors' work with a political agenda identified with environmental activism and the regulatory state. In practice, this means that advisors' research must be perceived either as neutral science (providing key information to influence new practices) or as a kind of moderate environmental politics based on "leadership" (guiding the industry through the perils of change for its own good).

As a consequence of these dynamics, one might expect that advisors would act in accordance with the cynical view of science and environmental politics, essentially acting as an appendage to the farm industry and conducting research that supports its short-term interests (as in Beck's view). After all, despite the structural and financial links advisors have to the local farm industry, their status as insiders is not granted automatically. On one hand, their local position in the valley helps to ensure them insider standing, but on the other hand, a strong association with environmental concerns and activism is one of the easiest ways to mark oneself as an outsider. Overall, in a situation where the boundary has been drawn this starkly, the simplest choice might be to position oneself on the safe side of this boundary, especially given the close relationship that advisors have with the farm industry (and their relative lack of contacts with campus-based UC researchers).

However, during my research with the farm advisors I found that their work—and the way they talked about it—was more complex than this. Advisors spent a great deal of their time working on environmental problems, testing and promoting new, "environmentally friendly" techniques of production to their grower clientele. Yet, they strove to temper any image of environmental activism by emphasizing that their work on these issues was in the long-term interests of the farm industry. They often justified this claim by contrasting their approach with the methods of the regulatory state. To better understand how advisors balanced their work with the farm industry and their attention to environmental issues, I asked them whether they considered themselves environmentalists. I expected that they might feel somewhat uneasy applying the term to themselves, given the kind of radical connotations often attached to it.[8] In fact, all the advisors were at pains to say that their environmentalist sensibilities were in balance with other considerations:

CRH: Do you think of yourself as an environmentalist . . . ?

GrapeAdv: I think those principles of being concerned about what you're doing—if you want to call that an environmentalist—I guess when you say environmentalist a lot of people would have a different view of what that may potentially be.

CRH: People out there chaining themselves to trees?

GrapeAdv: I'm not out there chaining myself to someone's spray rigs so they won't spray a pesticide. [CRH laughs.] That's what a lot of people have envisioned as environmentalists, that they're a bunch of crazies. But I think if your definition is someone who is concerned about what is happening . . . I guess yeah. We have a lot of words like that that mean different things to a lot of different people [laughs].

CRH: So, do you think of yourself as an environmentalist?

PlantPath: I suppose so, although those are loaded terms in the political world. I'm not an environmentalist in the true sense of the science. I'm a plant pathologist, and to me an environmentalist in the science world is more systems-oriented in their research. But if you are referring to the political term of someone who is promoting and defending the environment, certainly, I share those concerns and would put myself in that circle, in that broader sense. [By] the same token, I'm not really an activist—I choose not to be too involved politically and actively in terms of those issues.

In both cases, the advisors recognized themselves as environmentalists but were careful to avoid a definition of the term that was overtly political. Each framed his environmental conscience as an agenda that did not cross into protest or other forms of public statement. The key strategy for advisors is to frame their concerns about agriculture's environmental impact as balanced, sensitive to each side of an issue but giving full allegiance to neither:

GrapeAdv: Yeah, [environmental issues] are a big concern. I think we do get drawn in—sometimes, they can be somewhat political issues. But if you're drawn in, you're drawn in as kind of an unbiased source. Not too long ago [the county] had some concerns about [farming on] hillsides. And

[I got] contacted from growers and . . . from the [county] planning department. You get involved in—I guess you could call them political actions—but it's more because of what you know. Not because you're out there trying to promote one side or the other. I think most of us should stay pretty much neutral in those political type of issues.

Although both of these advisors ruled out protest or other radical measures, their concern about environmental issues is evident. It is clear that they had thought out the bounds of their concern, an appropriate set of possible responses, and how to sell these possibilities to the farm industry. In sum, advisors typically framed this process as leadership, subtly yet actively leading the farm industry toward more environmentally sustainable practices:

CRH: Can you say generally your sense of how environmentalism, the environmental movement, and environmental regulations from the state have had an impact on your job?

Entomol: That's a really good issue because it's a situation where you can either be a leader or you can just follow along. And I think if you're not willing to accept that environmental issues are important and that they are going to change the way agriculture is done, then you're just going to be following along and you're going to get left behind. I think, in this time, you have to realize that there are changes occurring and you have to help the growers deal with those. And you have to find a way to present those issues to the growers that makes them realize that this is something they have to deal with whether they like it or not. . . . As long as you explain to [growers] that you're doing this because you feel it's a future direction and something that's of importance instead of just saying, "What you're doing is wrong," I think it can be done very carefully. . . . And I think that the farm advisors should be responding to [environmental issues] and trying to be in a proactive mode rather than strictly a reactive mode.

This advisor described his leadership as both environmentalist and political; he implicitly invoked the role of the state and environmental regulation as a "future direction" of California agriculture. At the same time, he acknowledged that his own place in the ecology of farming required that this work be "done very carefully." He framed environmental change as an impending reality for which he could provide leadership, shepherding the industry through a period of turbulent political change.

In each of the previous statements, the advisors strove to balance their interest in addressing environmental problems with the interests of the farm industry. This is not to say that these interests can be neatly sorted out or that there is a consensus among, say, growers, on the definition of environmental problems and the appropriate response. But the overall definition of environmental problems and appropriate solutions called for an understanding of the larger farm ecology, an integration of local practice with moderate politics. Further, the advisors' attempts to maintain a balanced concern were themselves the consequence of a complex set of interests, perhaps the most important of which was preserving their local connections with and influence within the farm industry. In sum, this complexity troubles any overly simplistic view of how agricultural scientists conceptualize and respond to environmental problems related to farming.

Case 1: Nitrate Contamination of Groundwater

Examining some specific cases will help to demonstrate how advisors attempt to perform the kind of balanced leadership I described in the previous section. The first case concerns nitrate contamination of groundwater. California's vegetable industry grew up with the fertilizer industry, especially beginning in the 1930s and 1940s, when synthetic fertilizers became more widely available. At the time, Cooperative Extension farm advisors were among the greatest champions of this new technology, using field-based demonstration trials to show that for a modest cost per acre (relative to the potential sale price of the crop), growers could boost their yields and profits substantially. Vegetable growers in the Salinas Valley, whose crops are very expensive to produce but very profitable when commodity prices are high, promptly adopted the use of synthetic fertilizers. Further, as they gained more experience with fertilizers, they learned that overapplication does not harm vegetable crops but instead acts as a kind of crop insurance, maximizing yields for a slightly increased cost of inputs. Therefore, the treatment of vegetable crops with fertilizer, often several times per cropseason, quickly became a standard practice throughout the vegetable industry. This practice also became institutionalized in the fertilizer industry, which offers growers an application service, spraying fertilizer on the crops at fixed intervals throughout the growing season. The environmental downside to this practice is that any nitrogen or other nutrients that are not absorbed by the plants' roots may eventually leach into the water tables below the fields.

The solution to this problem seems deceptively simple: just get growers to use less fertilizer on their crops, and then less nitrate will leach into the valley's water supply. This, in fact, was the approach taken by a team of researchers working to address the nitrate problem in the late 1990s. The research team consisted of a Salinas Valley farm advisor, a UC extension specialist, and a few other researchers working with county-level administrative agencies. They were supported in part by grants from a statewide fertilizer industry association. The team developed and pushed for growers to begin using a "quick test" soil sampling system for monitoring their soil and making better judgments about the need for fertilizer at a given time. The quick test system is a relatively cheap and simple way to check the level of nitrate already present in the soil, thereby allowing growers to fertilize only when the plants require it, not on a fixed schedule regardless of need. In this way, the quick test was designed as new form of management that would serve as an alternative to the application schedules standardly offered by the fertilizer industry.

Based on field trials of the quick test system on several farms throughout the valley and other lettuce-growing regions of California, the researchers found that the system could be used to effectively reduce fertilizer use while maintaining the same yields. Despite the promise of the field tests, however, and the potential for growers to save a bit on fertilizer costs, the research team had a difficult time convincing growers to implement the quick test system in their own fields, and few growers ultimately adopted it. The quick test had run up against the standard practice of lettuce growers, which is to overfertilize as a kind of crop insurance; the research team could not guarantee that risk had been completely eliminated. As the following excerpt illustrates, people in the industry remained skeptical of the new approach, even after field trials using the quick test method had shown initial success:

CRH: It seems like, on [certain] problems, Extension can be pretty helpful, but are there other kinds of things where it's not always as clear whether [Cooperative Extension] is gonna be helpful or not?

Grower: Yeah, I think in some cases—Monterey County has a nitrate problem, and the state is rattling their saber about that. But it's a problem that does have to be cured and solved. Extension probably is not as capable of solving that type of problem because it's not a day or night situation. It's very long-term, because what you did two years ago probably still affects what you're doing today. And, it's very subtle, what your results are gonna be. And so they probably do a poorer job of that than they do on other stuff. [UC Specialist's] data shows that if you test [the soil] . . . you may be able to reduce how much nitrogen you put on by as much as 50 percent in some cases. . . . He's done this experiment two or three times, and it hasn't been accepted by the industry. Because nobody's gonna skip a $40 fertilizer application and possibly lose everything they got out there. They would rather make sure they have enough or too much. And, it's hard to sell that on paper. . . . Because everybody knows, well, you know: that was this time but what about next time? You don't get paid for blowing a $2,500 crop because you were trying to save $40. It's not their fault, that Extension isn't effective—the research has been good and the results have been proven. But it's subtle, and it's very difficult for Extension to do that well.

This grower described the quick test's results as "subtle," but what makes it subtle? For this grower and many others, the quick test represented an edge of change, marking off the institutionalized industry practice (overfertilize as insurance) and the threat of more drastic change (instigated by regulatory agencies). The subtlety stemmed from an uneasy sense of being placed on this edge without a clear and easy decision. One might argue that this grower and others were simply responding to an economic calculation about the risks and benefits of change, and it is true that the grower made a very direct reference to the costs saved by using the quick test compared with the costs of potential failure of a profitable crop. This decision, however, is highly context-dependent. The definition of the quick test as either a cursory act of maintenance or a more radical form of repair is shaped by the larger ecology of the farm industry as a whole, especially the grower's sense of risk in terms of regulation. In a sense, this grower was already thinking about the realities of the nitrate problem and even mentioned that the state was "rattling their saber," alluding to the threat of increased regulation. Further, the fertilizer industry itself had funded much of the research and development costs of the quick test system because of concerns about state regulation of fertilizer use. In all, the "subtleness" pointed to this grower's concerns about a changing farm ecology, where the risks of change and the risks of regulation made an uncertain balance between the institutionalized practices of fertilizing and the politics of environmental change.

The members of the research team understood that the quick test was on this edge of change, and this was actually their argument for the system's implementation. They argued that the quick test was a balanced approach that accounted for grower practice but also made changes that would appeal to regulators, perhaps even forestalling regulatory action. The following excerpt from an interview with the state specialist working on the quick test team illustrates the balance the team was trying to achieve and also the context-dependence of the choices:

UC Specialist: It's less compelling for a farmer to make changes in fertilizer use [with the quick test method] than if I was to tell them that they could change and increase production. I'm telling them that by monitoring their use of [nitrogen] they can save about $50 per acre on a crop that may cost around $2,000 per acre and yield much more. This is a lot different than telling them that there is a new irrigation method that could increase their yield by 15 percent. This would really make them pay attention. On the other hand, if the California EPA hauls [the industry] into court on a class-action suit, or starts taxing water use, or some other punitive method of dealing with nitrates, then I will suddenly be their best friend.

Here, the specialist pointed to the potential for more radical state regulatory action to change growers' attitudes toward the reduction of fertilizer use, and he also used this point when speaking to growers about the nitrate issue. In one meeting that I attended, the specialist explicitly raised the possibility of imminent state regulation as a reason that growers should begin managing their own fertilizer use before the state forced them to do so. This is similar to the advisor who talked about the need to "carefully" lead growers toward change. Although the specialist portrayed himself as potentially the industry's "best friend" if the state should decide to regulate fertilizer use, he also implied that he was already working in the best interests of the industry.

Ironically, although the specialist and the farm advisor working on the project both agreed that regulation would be the quickest way to change grower practices, neither of them was very confident in the ability of the state to implement a program that would effectively balance the needs of the environment and the farm industry. Just as growers expressed a wariness regarding the intentions of regulators and their potential misunderstanding of the larger ecology of agriculture, so the researchers declared similar worries:

Soil/Water: The state will probably step in pretty soon, because something like half of the groundwater in the southern part of the valley is unsafe to drink. The problem is, you can't just stop putting fertilizer on crops. It doesn't matter how clean your water is if you don't have any food to eat.

CRH: What about the reaction [to the initial success of the quick test system] from the regulatory agencies—they must be sitting up and saying, "We were right."

UC Specialist: They are excited by it, but I've tried to work with them to get them to see the realities of production ag. These agencies are mostly staffed with people who have backgrounds in biology or chemistry but have no experience or sense of the industry. They need to have an idea of how growers live their lives and do their work. The growers are not Luddites or [intentional] polluters. They have minimal profits and already have a lot of regulations and paperwork to deal with.

These research team members implied that an outright ban on fertilizers would be impractical. Once again, the researchers described their roles as leaders for change, but for balanced change. In addition, they talked about providing leadership to both the industry and the regulatory state, trying to make them see the "realities" of the full farm ecology. The research team hoped to use the "subtle" edge of change as a selling point to renegotiate the meaning of the quick test method as a reasonable and balanced mode of change. This process can be frustrating: at this stage in the negotiations over environmental change, the quick test did not satisfy the ideals of either the industry or the regulators.

When I conducted follow-up interviews with the Salinas Valley farm advisors in the spring of 2003, little had changed. Nitrate contamination of groundwater was still a problem in the valley, but other problems related to water and agriculture had moved to the fore and diminished the importance and urgency of the nitrate problem.[9] Whereas at the beginning of my research, both the regulatory agencies and the farm industry had been worried about lawsuits and negative publicity stemming from the nitrate contamination problem, these newer issues were now the focus of growers' and regulators' worries. As a result, growers were still showing little interest in the quick test system, and a grower survey conducted by a Monterey County water resources agency in 2001 showed that the quick test was rarely used as a management tool for regulating the use of fertilizers. Although 78 percent of respondents claimed that they used some method of testing to assess the amount of residual nitrogen in their soils, only 3 percent used the quick test. This suggested that most growers were using slower laboratory-based testing, most likely only once per season, instead of the continual management approach suggested by the quick test's promoters. Further, more than half the growers said that they used private consultants for advice on fertilizer use, including 50 percent of growers who reported relying on consultants from the fertilizer dealers themselves (MCWRA 2002).

In addition, the farm advisor who specialized in soil and water problems in the valley had resigned, and a new farm advisor had been hired to work on these issues. This new advisor had few hopes for the use of the quick test as a method of reducing fertilizer use, claiming that growers were "not comfortable with it" and felt it was too "low-tech" when compared with the sophisticated systems they used on their valuable crops. In this sense, the quick test system failed as an edge of change in at least two respects. First, it seemed too simplistic and potentially unreliable to growers, whose bottom line depended much more on the appearance and size of their crops than on the consequences of fertilizers' leaching into the valley's water supply. Second, the quick test no longer represented the cutting edge politically. The advisor and others concerned about groundwater contamination found it difficult to sell the quick test as a balance between overfertilization as crop insurance and more radical regulation from the state. In the end, the "subtle" choice faced by growers was eliminated through a shift in the definitions of local environmental problems, and maintenance of the status quo became an easy choice.

Case 2: Watching the Weather to Reduce Pesticide Use

The vegetable industry in the Salinas Valley has something of a reputation for very intensive use of several pesticides in a single application, in order to control multiple kinds of pests. Colloquially, this practice of mixing pesticides is called the Salinas Cocktail. Although, by law, growers need to have the recommendation of a Pest Control Advisor (PCA)[10] for use of a given chemical on a given crop, the actual application of pesticides often follows a program approach, as in the case of fertilizers, applied through a commercial pest management service. Therefore, the use of pesticides is also often treated as a kind of insurance, which can lead to overapplication and the possibility of pest resistance.[11]

The case I draw on here was a collaborative effort to control downy mildew (DM), an important disease of lettuce and other leafy vegetable crops, while reducing pesticide use. As with nitrate management, an improved management regime is one alternative to an outright ban on pesticides used for DM. DM is a foliar (leaf-based) disease most prevalent in cooler, more humid coastal vegetable-growing areas. This is in contrast to drier growing areas, in Southern California and Arizona, where vegetables are grown during the winter months. In these desert regions, DM is never a problem. Even within the coastal growing areas, however, the incidence of DM can be quite variable; sometimes the disease is rare and can be controlled, while at other times it is in every field and does great damage. Therefore, a program approach, when used to control DM, can be wasteful because this method sprays for the disease whether the risk is small or large.

In contrast to the relatively low-tech quick test system, a group of university researchers (including the plant pathology advisor), the Lettuce Board, and equipment manufacturers were, at the time of my research, working on a rather high-tech solution for managing DM (and the chemicals used to treat it) on lettuce. The solution involved the use of electronic weather stations placed throughout the vegetable-growing areas of the county. These weather stations monitored climatic conditions in their vicinity and radioed these data to a central computer system that processed them using a special DM algorithm, a model developed to predict, based on climatic factors, when DM was likely to be a problem on lettuce. This method of predicting DM would then allow growers to fine-tune their application of pesticides, spraying for DM only when necessary. In one of my earliest interviews with the farm advisors, in the summer of 1997, the plant pathology advisor described the DM weather station project and some initial results:

PlantPath: We're trying to control diseases, in the most economic way, and also [in a] way that is less disruptive to the environment. So, [DM] for example . . . is an important foliar disease on lettuce. We know we can spray for it; we're trying to find ways to spray less for it, so that we can conserve chemicals and introduce less of [them] to the environment. That covers [several] priorities: it's an economic concern, it's a high priority for the industry, [and] there's environmental issues.

CRH: Is that a project that you're working on now?

PlantPath: Mm hmm. It's kind of an interesting project. . . . We have these new . . . weather stations that go out into the field. And, by radio telemetry, they radio . . . to us all these environmental parameters. We have a model, that based on those parameters, can predict whether disease will come or not. And if the weather station beams the data and the computer processes it and says, yes there'll be downy mildew this week, we'll tell the grower to go spray. If it says no, it's too warm or too dry or whatever, don't spray, we'll tell the grower, don't spray, your risk is light. And we want to, again, conserve chemicals and reduce that input. So we're trying to test that system right now.

CRH: How's it working so far?

PlantPath: Uh, so far ambiguous—you know, a couple cases looked promising, and others, it looked like it flopped. The model was developed at Davis and worked well in [the laboratory]. . . . So, that's a good start but we want to now implement it here and see if it works commercially. So far we're not real happy with it. We think a better-proved model is probably called for, but we're still in the middle of testing that.

Note that the plant pathology advisor emphasized, at the outset, the importance of linking environmental and economic concerns. DM was a good problem to work on because it was important to the industry (caused a lot of economic damage) and required a lot of pesticides (caused potential environmental damage). The advisor also pointed to some of the varied groups involved in developing this system, of which there were many. Finally, he mentioned that the initial use of the system had given mixed results, and that the model needed to be further refined before the full system was commercially viable. Over the course of the next few years, I continued to collect data on the progress of this system, both from the plant pathology advisor and from other informants who were involved or at least familiar with the project. In all, the factors mentioned in this first interview excerpt—the important connection between economics and environment, the diverse coalition of actors involved, and difficulties with the system's implementation—combined to cause conflict and controversy over the DM project.

I spoke briefly with the director of the Lettuce Board about the DM project, and he emphasized the many different people working on the system: Lettuce Board members, university researchers, and equipment manufacturers. He told me that this mix was unusual because these different groups have various interests and would not usually cooperate. My research, however, shows that these various and divergent interests were the most important elements of the conflict involved in this project. Because these interests played such a large part in the story, it is important to identify and detail the complex cast of characters. I have already mentioned that the plant pathology advisor (PlantPath) was involved in the project, but the primary researcher based at Davis (DavisSci) developed the computer model to predict DM from climatic factors. A former UC scientist (IndustrySci) left the university for a job with a private company that wanted to market the weather stations commercially (AgCo). Finally, there was the Lettuce Board, which provided funding to both the plant pathology advisor and the Davis researcher to develop and test the DM prediction system. The Lettuce Board's director as well as a subcommittee of industry representatives oversaw the implementation of the system.

The initial source of tension within this project came from a conflict between the Davis researcher and the Lettuce Board. After the researcher developed an initial model for DM prediction and found some positive results in the laboratory, the Lettuce Board wanted to begin field-testing the system. The researches believed that further campus-based tests were required before field trials could begin. Eventually, after members of the Lettuce Board had asked her several times to field-test the model and were refused, the Lettuce Board discontinued her funding and asked the plant pathology advisor to field-test the system for them. In an interview with another advisor who was *not* involved in the DM project, this falling-out came up in our conversation. This advisor had been speaking generally about his relationships with commodity boards, and I include an excerpt here because it shows the power of commodity boards to shape research priorities. This interview took place in the fall of 1997, about three months after my initial conversation with the plant pathology advisor about the DM system:

Advisor: If you have a history of delivering on what you said you'll do, then [the commodity boards] tend to pretty much back off and let you do that the way you feel you need to do it. If you have a history of proposing to do something and then you're not coming up with an answer, or coming up with something that they don't feel is useful, something they don't understand, then they're gonna find a way so that it's more workable for them.

CRH: Is that what you were saying before, about, how when you first [started working for Cooperative Extension], you had to let the Lettuce Board kind of get to know you?

Advisor: Yeah, that's exactly why I said it.

CRH: And you develop a reputation as someone who can do something without screwing it up or whatever?

Advisor: Right. And I will ask and answer questions that are appropriate to them in the field. An example of something that's happened recently with the Lettuce Board, is that . . . they gave a grant to [DavisSci] from Davis to do some modeling work for downy mildew. You might want to talk to [PlantPath] about this because he's getting involved with this and it's kind of an interesting situation. Basically, she was doing the very academic sort of [modeling] . . . and [the Lettuce Board] tried on numerous occasions to affect the kind of work that she was doing. And they weren't really successful with that. So they just . . . stopped funding her. [The Lettuce Board] tried [to say], "This is what we want" or, "This is what you're doing and we're not happy with that; you need to do *this*." And when she wasn't willing to do that for them, they said, "OK, fine, you're not doing any more research for us." They shut it off. And then they go to someone like [PlantPath] and say, "OK, now [PlantPath] will *you* do this for us?" So it kind of put him in an awkward position.

Like the advisor in this interview, the plant pathology advisor had a reputation for providing solutions to local problems and often worked very closely with the Lettuce Board and other commodity boards on these issues. In this case, however, the Lettuce Board was not giving him a grant for a project designed and implemented according to his own ideas, with the kind of freedom from interference described in the previous excerpt. After their difficulties with the Davis researcher, the Lettuce Board wanted specific assistance from the plant pathology advisor. The history of this project and the way that he was brought in to salvage the project made him uncomfortable with the situation. In another interview, in the early spring of 1998, he was even more reserved in his opinion of the DM system and frustrated by the Lettuce Board's and the company's push to implement it. In this excerpt, he described the many groups at play in the DM project and their tendency to have different interests from his own:

CRH: I was wondering if you could talk about what you feel are some particularly difficult problems that you have been able to solve—or maybe even some that you feel like you really couldn't solve because they were especially difficult.

PlantPath: OK, . . . how about one that isn't solved yet?

CRH: Yeah.

PlantPath: Because the biggest one that I can think of . . . [is] the prediction system for downy mildew on lettuce. . . . [DM is] the most important disease that the Lettuce Board has identified for . . . [lettuce]. So there's industry interest. There's lots of funding, because they want this done. And so [I] jumped in because of that. It's been difficult for several reasons. One: technically it's a very difficult issue or system. If you look at all the literature on [these systems] there's not a whole lot of these systems that work anywhere in the world.

CRH: Computer models, you mean?

PlantPath: Right, where you have a computer model, weather data, and whatever equipment that [use a] model to predict when the disease is gonna come. The track record is pretty bad—there's nothing that really works that well consistently. So it's a challenging technical area. I think on top of that, I really feel my limitations in this project, in terms of my own background. Because I'm not a modeler—I haven't worked with modeling systems. . . . This is my first try at it. The project is well within my scope, because they're not telling me to devise a model. That's been done—they want me to test it. So that's pretty straightforward, but I feel frustrated because I'd like to be able to look at the data . . . and say, with my expertise, "If we change this part of it, this will really make it much better. . . ." But I can't do that—I don't have the technical expertise for that. So I'm basically just running the model and seeing what it does, being able to say, yes or no, it's working well. So I find it frustrating, because, if I could add more to that, I think I could take it further. But I can't do that. . . .

And the last thing on this particular example: I think it's the *only* project I can think of, in my brief time here, where it's not been real smooth working with all the cooperators. I'd say, with the exception of [the DM] project, *everyone* else, from farmer [to] field person [to] PCA, they've really been great in cooperating with us, and I can say nothing but good about the working relationships in these other projects. This one is rather different in that it's not been that smooth. Some of the companies that provide the hardware and stuff like that—they're trying to sell units. So when we come out and say we're not sure it's working well, there's some conflict there. Because they're saying, "It's working well. And we want to sell 'em to you, Mr. Grower." And I come out and they ask me, "Well we hear it's working well." And I'll say, "Well, I'm not really sure it is." And that creates hard feelings, see. Because we're not supporting what the commercial people are saying.

CRH: Well, the Lettuce Board wants to [implement the system] now, right? That must be frustrating too.

PlantPath: Yeah, that's frustrating, because the board . . . well, there's a subcommittee that's pushing for that. So the board as a whole says, "You subcommittee [members], you meet with [PlantPath] and come up with a recommendation." And the subcommittee is very much in favor of pushing ahead faster than I want to. So, I think I've been pretty good about my response. I'm encouraging them and saying, "I want to be supportive of the effort." I've said, "If you need me, I will still provide some technical support." Which they want—they want Extension involvement in it.

Once the plant pathology advisor had collected data from the field trials, though, his findings could be interpreted in multiple ways. In fact, representatives of the company that intended to commercially market the DM weather station service began citing his data to show the potential benefits of the system. In one case, the advisor was out of the country at a conference when a company scientist (a former UC researcher) presented a very positive spin on the data before an audience at the Lettuce Board's annual research meeting. In another instance, he was in the audience when the company scientist discussed the DM system at a national plant pathology conference, claiming that the system had already increased the productivity of lettuce growers in the Salinas Valley. Although the plant pathology advisor was tempted to "make a scene" and challenge the scientist's presentation, he maintained his diplomatic stance. In an interview he expressed irritation toward this appropriation of his data yet still tempered his anger, saying, "It's annoying—that's the best thing to say about it."

When I last spoke to the plant pathology advisor about the DM system, in the spring of 1999, he was no longer involved with the DM project at all, and the company was actively marketing the weather station service to lettuce growers. The advisor had resigned himself to their promotion of the system, claiming that he understood the self-interest behind their positon. Ironically, this extension advisor was more concerned about the inflated claims being made to researchers at a national conference than about the kind of marketing being directed toward growers. He believed that, unlike the researchers, the growers had the local knowledge to decide for themselves whether the DM system was effective.

PlantPath: If a few years from now a lot of people are using it, then we will know it works. If no one is using it, then we'll know it doesn't. We have the best lettuce growers in the world here. [AgCo] isn't going to fool them.

The advisor's relations with the Lettuce Board, however, were a little more tricky. He told me that members of the board had been pressuring him to be more supportive of the system, especially because its successful implementation would be a useful showpiece, allowing the industry to show regulatory agencies that vegetable growers had been proactive in their attempts to better manage pesticide use. Again, he emphasized the care he had taken in positioning himself: he still diplomatically explained to growers that the system "needs more work" and had not given into the Lettuce Board's pressure for a cheerful endorsement. In the end, he was able to distance himself from the DM project, although he could not afford the kind of stubborn attitude that the Davis researcher had shown in disagreements with the Lettuce Board. The researcher could survive the ill will of the Lettuce Board and find other funding sources for her campus-based research, but the advisor could not afford to alienate the board and needed to treat the conflict much more delicately.

This case is instructive for exploring the tensions that can surface in collaborative projects to minimize the environmental impact of agriculture. In addition, the DM case points to the complex politics surrounding science and the repair of environmental problems associated with the farm industry. What is really being repaired through the DM project? It is interesting to note which actors were most in danger of becoming outsiders in the eyes of the industry, and for what reasons. Although the boundary between inside and outside is often drawn by marking off science from politics, both the Davis researcher and the plant pathology advisor risked outsider status by taking a very scientistic approach to the DM project. The Lettuce Board grew impatient with the strict scientific standards of the Davis researcher. The advisor needed to walk a fine line to avoid a similar fate and was frustrated by the board's unwillingness to further test and refine the model. Thus, the farm industry's attempt to send a message to regulators actually interfered with and hindered these researchers' attempts to improve environmental practices in agriculture.

The Balance of Power and Social Change

Any serious attempts to repair agriculture's environmental impacts imply changes to the fundamental practices of farming. Chemical inputs used to fertilize crops and control pests are an integral part of modern vegetable farming; it is hard to imagine the scale of production in a place like the Salinas Valley without them.[12] As I have argued throughout this book, analyzing the link between practice and power is essential for understanding how repair works (or not) in industrial agriculture. The cases I present in this chapter raise questions about the character of this repair. What is being repaired? Can farm advisors really do much to repair environmental problems in agriculture?

Although the quick test and the weather stations were quite different in their degree of technological sophistication, they shared a focus on management of existing practices. In this respect, each solution was an attempt to maintain the existing system of practice for vegetable farming, a way to show that the farm industry can effectively regulate itself without having to take on too much change. Neither case was an attempt to "greenwash" the industry's environmental impacts (not entirely, anyway) (Athanasiou 1996; Austin 2002). At the same time, neither case was an especially radical transformation of the industry's farming practices. Similar to the field trials used to create consent around a new practice (see chapter 5), the quick test system and the weather stations were conservative, middle-of-the-road approaches to environmental change. Thus, the object of repair in these projects was industry control and power just as much as the environment itself.

The complex interaction between advisors and growers vis-à-vis environmental issues bears upon the question of how much impact applied scientists can have on environmental change. The incrementalist approach that advisors take to repairing environmental problems reflects the location of the advisors themselves: their integration into the local community of growers gives them a special entrée that outsiders do not have and cannot easily acquire. Because they are situated in this space, familiar with the local farm community and its systems of production, advisors have the potential to create networks and devise new techniques that may not be apparent to either the industry or to regulators. They have "gone local" and can "see" the ecology of the farm industry in ways that others may not. This location, however, also has special pressures, including the temptation to choose an easy compliance with growers' desires to avoid change. No doubt the advisors' wariness when describing their identities as environmentalists and their distrust of regulators has at least some basis in their very localized domain of influence. At the same time, advisors form close connections with some of the biggest contributors to water pollution, providing the advisors with a unique chance to create the conditions for environmental change. If they can make even a relatively modest change in the practices of the industrial growers who provide a majority of the produce for the United States, then their work could have a larger overall impact.

7 Conclusion: The Future of Agriculture in the United States

A Great Hollowing

In chapter 2, I described the growing pains of Cooperative Extension in California, expressed especially through the reports of its director, B. H. Crocheron, during the 1920s and 1930s. Crocheron wrote of his wish to "unscramble" the agriculture of that time, consolidating those farms he deemed inefficient into larger and more rational units. Crocheron got his wish. U.S. agriculture in the twenty-first century is a rationalized and consolidated system of food production; the "mess" of earlier times has been replaced with a context where fewer than 2 percent of Americans now work in agriculture, and yet our food production leads the world and is unrivaled in the history of humanity. Ironically, however, the same processes that began this transformation during Crocheron's time are still at work today, and the remaining farmers face several of the same problems that were prevalent at the turn of the twentieth century. While the number of Americans working in agriculture continues to shrink, and farm sizes continue to increase, the definition of who is a small or marginal grower remains in flux (Hallberg 2001). The slogan "get big or get out" defines farming in many sectors, and the result has been not the preservation of farming communities but instead a great hollowing.

As I write this conclusion, I make my home in central New York State, an agricultural region far different from the Salinas Valley. Whereas the acreage of farmland is about the same in Monterey County and the county where I live, the corn and dairy farming of this area produce dollar revenues in the tens of millions each year, nowhere near the scale and intensity of California's multibillion-dollar vegetable industry. Furthermore, this farming region is experiencing a long-term decline, and I can drive on country roads in any direction from my home to see abandoned farms marked by crumbling barns and fields filled with dandelions. Whereas once dairy farms in this region could survive with a herd of 50 or fewer cows, that number has steadily crept upward to over 100 cows in 2006. More than half of New York State's dairy farms failed between 1980 and 2000; the remaining farms are increasingly employing a set of industrialized institutions, practices, and technologies first pioneered in the nation's leading dairy state, California. Driving through the dairy regions of California, especially the huge operations typical of the Central Valley, where herd sizes average more than 900 cows and sprawling feed lots often contain thousands, provides a vision of the likely future of dairying in places like New York State.[1]

These landscapes raise difficult questions about the future of farming in the United States. Without trying to return to an idyllic agrarian vision, one can ask whether that kind of food production is what our nation wants and needs. Considering the steady decline of farming as an occupation and way of life during the last two centuries, there seems little possibility of turning back the economic and technological trends that drove these changes. And yet, one of the greatest insights of the academic field that most inspired this book—science and technology studies—is that ideas and things that we take for granted have a history of contingency, that there are many parallel universes that might have been and perhaps still could be. Recent years have seen a surge of interest in issues related to the politics of food and agriculture, suggesting the possibility of renewed debate and fresh ideas regarding the future of U.S. farming.[2]

In this chapter I offer a brief critique of the economic and technological determinism that governs much of our thinking about agriculture and inhibits optimism about change. My intention is not to make a comprehensive set of proposals about what the future of agriculture ought to be, but I do argue that U.S. agriculture at the start of the twenty-first century is not environmentally or socially sustainable; it needs transformative repair. Further, and perhaps in conflict with the apparent logic and history of prior agricultural change, I believe that the land-grant universities, and especially Cooperative Extension, have an important role to play in this transformation.

The Contingency of Economic and Technological Change in Agriculture

Some of the best and most vivid descriptions of the role of economics and technology in agriculture come from novelists who sought to understand the extensive changes to U.S. agriculture in the early twentieth century, such as Frank Norris and John Steinbeck. Norris (1901) describes the Union Pacific Railroad monopoly as a giant "octopus," and Steinbeck (1939) portrays banking interests as a "monster." Each depicts a profound dilemma in the economics of farming: although the economy is a human creation, it has somehow taken on a life of its own. For example, in this scene from Steinbeck's *The Grapes of Wrath*, bank representatives have come to foreclose on the land of a group of tenant farmers, and the conversation between these two groups reveals the monster logic of economics:

The tenant men looked up alarmed. But what'll happen to us? How'll we eat?

You'll have to get off the land. The plows'll go through the dooryard.

And now the squatting men stood up angrily. Grampa took up the land, and he had to kill the Indians and drive them away. And Pa was born here, and he killed weeds and snakes. Then a bad year came and he had to borrow a little money. An' we was born here. There in the door—our children born here. And Pa had to borrow money. The bank owned the land then, but we stayed and got a little bit of what we raised.

We know that—all that. It's not us, it's the bank. A bank isn't like a man. . . . That's the monster.

Sure, cried the tenant men, but it's our land. We measured it and broke it up. We were born on it, and we got killed on it, died on it. Even if it's no good, it's still ours. That's what makes it ours—being born on it, working it, dying on it. That makes ownership, not papers with numbers on it.

We're sorry. It's not us. It's the monster. The bank isn't like a man.

Yes, but the bank is only made of men.

No, you're wrong there—quite wrong there. The bank is something else than men. It happens that every man in a bank hates what the bank does, and yet the bank does it. The bank is something more than men, I tell you. It's the monster. Men made it, but they can't control it. (43)

The economic monster is unstoppable in this example, making it hard to consider or even conceive of alternative visions of farming and land tenure despite the apparent failure of the then-current system. In addition, these narratives of economics out of control are typically tied to technologies and technological change. In Steinbeck's story the monster is represented most graphically by the tractor that literally drives through the tenants' door and pushes them off their land; Norris's octopus is the nineteenth-century railroad system.

These combined forces—economics and technology—form a powerful discourse about the inevitability of one kind of agricultural future and the impossibility of others. Naturalizing these institutions allows actors to portray change as blocked by an economic and technological dilemma that restricts their agency and deflects it to other social groups or to forces outside the control of society altogether. Over the course of my research, in talking with many growers and researchers, I often heard similar versions of this dilemma, especially when they tried to defend niche market farming against charges of environmental degradation and calls for change. Most of my informants agreed that environmental problems are important and that changes are needed. On the details of this change, however, the answers were equivocal, and researchers and growers often invoked this dilemma to defend existing practices and structures. For instance, a researcher working in the Department of Vegetable Crops at UC Davis narrated this version of the dilemma:

Researcher: It's easy to sit back and say, "Well, gee, why are they spraying all those pesticides?" or "Why are they plowing up all that land?" or whatever. It's more difficult to sit down and say, "OK, if we don't do that, what's the consequence?" or "If we do it a different way, what's the consequence?" It can be done in various ways. . . . There are alternatives—if we're going to change some of these things, that's fine, but what do you have to do to allow that to happen? You're going to have to change your consumer habits, you're gonna have to be willing to eat differently—eat less meat or be willing to eat fruit with spots on it. . . .

There are lots of things that can be done, but, in a sense the farmer per se is a dependent variable in this, not an independent variable. . . . The farmer is dependent upon what people are willing to buy, what the market is willing to pay, [and] what [farmers] have to pay for their inputs. And so they're always just making an economic decision [on] how to survive. If you really want to change the system, change the demands, change the consumer preferences, change the rewards and punishments.

Here, the farm industry is depicted as just one part of a larger system of capitalist production. Farmers choose their practices and technologies based on economic criteria, and therefore production largely depends on the tastes and preferences of consumers. Because of this dependence, growers need to pass on to the consumer any extra costs associated with mitigating environmental damage, or go out of business. Consumers represent the monster, and they are in control.

A corollary of the dilemma invokes the size and complexity of this food system and suggests that conventional farming systems are much more economically efficient on a large scale for feeding the billions of people of the world. In this example, a grower warns of the danger of abandoning conventional methods of production:

Grower: They're been farming the United States for 250 years—I mean, how sustainable do you want it to get? . . . I happen to be old enough to remember what organic really was. Damn near was. I remember when DDT came out in the '40s, it was a miracle. We found tomatoes without bugs in 'em. Now it killed birds, and that's unfortunate and I'm glad that we don't have it anymore. It might have caused cancer, and I'm certainly glad we don't have it anymore if that's true. But . . . I can remember, [as an] elementary school student picking tomato horn worms off of the tomato plants because that was the only way you could kill 'em. You picked 'em off, threw 'em on the ground and stomped on 'em. That killed 'em. It's very ineffective. You sure as hell don't feed 250 million people or a billion people by picking tomato horn worms off by hand.

This version of the dilemma raises the specter of mass starvation, with inefficient organic methods of production causing food shortages. Agriculture is again depicted as a dependent variable; the rest of the world is critically reliant on conventional agriculture and its advantages of productivity, scale, and technologies; and the hero at the heart of this cautionary tale is a pesticide. This grower emphasizes both the structured weakness of growers and their responsibility for feeding the world. He implicitly admits that the system needs repair but claims that a truly transformative act of repair would be dangerous, perhaps even impossible.

These narratives of economic and technological determinism are clear examples of discursive repair, a means for justifying and maintaining the logic of a system of production. It would be a mistake to underestimate the power of these discourses for regulating perceptions and shaping the practical and institutional structure of production itself. Recent social science research on the role of economic discourses, in particular, shows that even abstract economic theories create a logic and a symbolic context that can have a "market effect" of their own.[3] And yet, many of the examples I described in previous chapters point to the limits of economic and technological imperatives. For example, in chapter 4, I described the Spreckels Sugar Company's attempts to encourage new technologies that could reduce or eliminate the use of migrant labor for thinning and harvesting sugar beets. The continuation of the Bracero Program until the mid-1960s gave growers a cheap and pliable source of labor that proved hard to give up. Although Spreckels could marshall economic, political, and even moral arguments to support its claims for the superiority of mechanized beet production, use of migrant hand labor was a long-established and seemingly vital element in growers' control over labor relations and practices. Similarly, in chapter 6, I described contemporary attempts by farm advisors to reduce the environmental impacts of intensive vegetable production in the Salinas Valley; growers clearly did not always choose their practices and technologies in terms of strict economic rationality. In fact, in the case of the weather stations implemented to track the threat of downy mildew and help growers make better choices about use of pesticides, the farm industry invested a considerable amount of resources in them in spite of their ineffectiveness.

Although one of my interview subjects claimed, "Any way that [growers] can use less pesticides, less inputs, they're gonna be happy because they're saving money," the effect of economic and technological incentives is actually contingent on a web of factors stemming from the place, practices, and institutions that structure how growers produce their crops. The common theme in these examples is the role of existing structures, including everything from the routine details of how growers manage to conjure crop after crop from a small valley to the wealth that allows them to influence the research priorities of agricultural scientists and the legislative agenda of the state. The ecological approach I have taken in this book brings these diverse factors together in order to clarify the structures and motivations of actors working in a complex system of food production. In addition, while my focus has been on agriculture here, the concepts of repair and ecology are applicable to other settings, especially those where actors have invested considerable resources in a particular form of produc-

tion, and control over the places, practices, and institutions of this production yields capital and power.[4]

Cooperative Extension and Public Support of Agricultural Research

Nearly every year, when the U.S. Congress debates large appropriations
bills, or when a new farm bill comes up for funding, public attention is
drawn to the role of farm subsidies in U.S. agriculture. News stories about
millions of dollars to fund the storage of peanuts, to support tobacco
growers, or to finance research on the "sex lives" of insect pests provide a
convenient way for journalists to manufacture controversy and for politicians to portray themselves as fiscal conservatives. The very terms used to
describe appropriations for a specific project in a particular location—*pork*
and *earmarks*—come to us via the farm. Agricultural subsidies are a complex
set of economic and political structures, and a full discussion of them is
beyond the scope of this book. But, at base, subsidies are a method of
repair: they are meant as a financial or technical inducement to change
grower practices and reorient their farm ecology. As such, it makes sense
to consider Cooperative Extension itself as a kind of farm subsidy and to
examine the logic of funding a nationwide system of agricultural expertise
when fewer than 2 percent of Americans continue to farm. As more and
more people have few or no connections to farming, who still believes
that farming is a public good that needs science for protection and improvement? In short, who still thinks that farming needs state-sponsored
repair?

During my conversations with the farm advisors in Monterey County,
we often talked about the privatization of applied agricultural research.
The advisors felt they had good relations with the local farm industry
and could expect continued support from growers, but they were less
optimistic about the budgetary priorities of the UC and the county. In
fact, there are several factors that point to the increasing privatization of
advising work and the likelihood that this trend will continue. First,
though private, for-profit sources of advice in farming have been around
for a long time, the emergence of Pest Control Advisors (PCAs) in
California and other new professions associated with agricultural expertise has had an especially strong impact on the move toward privatization

of farm advising. PCAs are now largely responsible for the day-to-day advice that growers receive about particular pest control problems, what types of controls are available, and which controls to use.[5] Although this kind of face-to-face interaction between experts and growers has not been privatized to the extent cited in some reports (at least in California), the trend toward increasing privatization is clear (Wolf 1998; 2006; Warner 2007, ch. 4).

Second, much of the research that is still done by UC advisors is funded through private grants from commodity boards or chemical and equipment manufacturers. These private sources of funding have a very strong influence on the kind of work that advisors do, but the availability of private funding also has wider implications for the future of Cooperative Extension. For example, the UC is increasingly asking growers to foot the bill for the salaries and facilities of researchers, items that in the past were provided through UC and county budgets. The ability and willingness of growers to underwrite a larger share of the expenses for agricultural science could entice the UC to continue exploiting this source of support, especially as budget pressures force it to retrench while still meeting the demand for new areas of research.

Third, several western European countries, Australia, and New Zealand have already implemented more fully privatized systems of university-based agricultural extension, and the UC is moving toward this model itself. Most of the funding for advisors' research comes from private sources, and PCAs provide much of the daily advice to growers about pest control. In this respect, advisors are already engaged in the money hunt that all university researchers must participate in to support their work. If the UC asks the industry to provide the salaries and facilities of more researchers, it will be hard to distinguish this system from the fully privatized extension systems of other countries.

Given these trends, privatization is already under way, and farm advisors expressed strong reservations about this process. They speculated that if extension work became fully privatized, most of their time would be spent hunting for financial support, and serving smaller, less financially secure growers would be even harder than it is now. For example, the director of UC Cooperative Extension in Monterey County voiced concerns about the potential for privatized agricultural research to limit access to new knowledge about food production:

Director: The big companies . . . would just hire [the advisors]—if we weren't here, I really have the feeling that [the big growers] would hire somebody. If they need it done they'd find somebody. They'd hire [Entomol] or they'd hire [PlantPath]. [big sigh] Um . . . but then . . . some people would say . . . OK, let's say the taxpayer says, "Well, why don't you just let [the growers hire the farm advisors]?" Well . . . [when] Cooperative Extension was started, the government said it's important strategically and politically and economically for people to have abundant and secure food. And, now we've gotten so good at that that people take it for granted. But if there was only large companies that [were] farming and doing the research, then they could control the food, too. And we wouldn't want food to become a monopoly. So, there's reason for the research to be, you know, unbiased and open to anyone.

The director's intuition that if the UC ended its Cooperative Extension program altogether, big growers would simply hire the advisors for their own needs is probably correct. In fact, many of the growers I interviewed emphasized their companies' research needs and said that they would find someone to do this work if the UC could not provide it. At the same time, growers often expressed a preference for advisors and their research to remain somewhat independent from any one produce or private consulting company. This grower, for instance, noted that advisors helped support the credibility of her company's claims in disputes with customers:

Grower: We . . . work very closely with [PlantPath], with [Entomol], with [Soil/Water], you name it—we use them to help us on diagnosis, problems, et cetera. Because, in a situation where we're selling our plants, it's always nice to be able to use the farm advisors as an outside, independent, nonbiased source of information. If we say, "[PlantPath] diagnosed this [problem] as blah blah," the [customer] is definitely going to believe it, instead of having an in-house person say, "This is what the problem is." Having that outside source is very valuable, in terms of, you know, everybody trusts the farm advisors and everybody knows that they're very credible. So, we use 'em a lot for those kinds of things.

Given these preferences, the farm industry will likely lobby very strongly for continued public funding, and even expansion, of Cooperative Extension. Along with the substantial savings in cost for the industry, having a public source of information and research on farm problems lends a sense of stability to an unpredictable and competitive enterprise.

Despite advisors' and growers' interests in maintaining state-funded sources of local agricultural expertise, it is unclear whether and for how long the political will for public support of these resources will continue. U.S. agriculture has come so far from the Jeffersonian vision of a rural farming class—can one still argue for the idea of agriculture as a public good? Because food itself is a public good, I believe that agriculture remains one as well and that the communities and agricultural environments where food is produced are of particular importance for the future of the United States. While it may be unrealistic to expect that farming can or should be turned back wholly to older modes of production, critics of industrial agriculture such as Thomas Lyson make a compelling case for a return to more localized systems of food production and distribution, a model Lyson terms "civic agriculture" (2004). Local community-based problem solving is a key element of this model, but rather than seeing a role for the land-grant universities and Cooperative Extension, Lyson focuses on the failures of these institutions to promote the vitality of agricultural communities in the past. There is certainly plenty to fault in this history, but alternative visions of local collaborations between growers and experts—in many cases actually inspired by extension experiences in the developing world—suggest the possibility of more holistic and egalitarian modes of cooperation (Chambers, Pacey, and Thrupp 1989).

The environmental impacts of industrial agriculture are another source of concern for the future of U.S. food production. As I described in chapter 6, growers and agricultural researchers face difficult questions of how to maintain industrial agriculture in its current form and scale, especially its use of synthetic pesticides and fertilizers, its reliance on fossil fuels, and the ever-increasing scale of food production. Further, if predictions about the future of the Earth's climate in the twenty-first century hold true, then human-induced climate change will likely have strong impacts on U.S. agriculture. Recent projections indicate that weather will become more extreme, with long periods of drought and intense rain patterns that may increase the risk of flooding (Easterling et al. 2007). These changes will disrupt the farming places as well as the knowledge and techniques that growers have developed to grow crops in these locales. If the Nation's Breadbasket becomes too dry or the Great American Salad Bowl is perpetually flooded, who is going to develop solutions to these place-bound changes? Publicly funded advisors are in a unique position to develop farming techniques that prioritize the protection of agricultural environments and to use their local credibility to act as leaders for environmental change.

If Cooperative Extension survives in the twenty-first century, it will likely do so through the same negotiations over its mission that have taken place since the formation of the land-grant system in 1862. These negotiations point to the limits facing farm advisors as agents of change. Throughout this book I have emphasized that the kind of local expertise employed by farm advisors is most suitable for promoting modest change, or repair as maintenance. What would it take to *transform* U.S. agriculture? Several of the examples presented in earlier chapters suggest that local politics are not necessarily conducive to transformative repair, especially given the structures of power that local elites shape and control. Instead, state policies and regulation are the most likely source of wider change. In an era where the term *Big Government* is not a winning discourse, state and national politics are nevertheless an essential venue for promoting change in agriculture. When the state weighs in on farm labor practices or environmental protection, growers and researchers listen. While the wealthiest growers have shown considerable influence over the political process themselves, the very consolidation of U.S. agriculture has paradoxically limited the voice of farm interests, especially compared with their historic power.

A key factor in the success of legislative and regulatory change for agriculture are the voices of scholars and intellectuals providing new ideas and pushing for alternative visions of food production. Although this adds yet another layer of expertise and interests to an already complex ecology of power, knowledge of how agricultural systems work may be used to understand the potential constraints on change as well as to propose change. In particular, I believe that the ethnographic approach I have taken here provides a useful method for modeling order and change. By going to local places and investigating the connections between place, practice, and power, ethnographers can learn about these local ecologies, write and teach about them, and reimagine the links and flows. U.S. agriculture will continue to change—the true dilemma is whether in the future we can balance the interests and practices of farm communities, consumers, and agricultural ecologies.

Appendix: Methodology

The data for this book were collected through a combination of methods, including participant-observation fieldwork, semistructured interviews, and analysis of historical documents. I chose this mix of methodologies because my theoretical interests, especially the ecological relationships between place, practice, institutions, and power, require very detailed attention to both the flow of practical activities on the ground level and the larger context that shapes and is shaped by this practice. Central to my choice of these methods are two assumptions. First, practice cannot be studied solely through what people say; one must also see what they really do. Therefore, I chose a site for fieldwork where I could actively participate and observe the activities and interactions that my research subjects had with their work and each other. This choice allowed me to see things about extension work that I could not have read from interview transcripts or the kind of historical documents I collected. For example, in chapter 5, I describe advisors' use of field trials to produce consent around new farm practices and technologies. Although I learned a great deal about field trials from my interviews with advisors and growers, fieldwork gave me insights into aspects of the trials that did not come through in these discussions, largely because many aspects of practice often remain transparent to the actors themselves.

Second, just as I chose participant observation to account for some of the limitations of oral and text-based sources of data, I chose to use these latter methods to expand the study of practice beyond specific, local actions. This situational context shapes local activities and relationships, and so it counts for more than just a kind of background. Therefore, I have used data from interviews and historical documents to inform both local practices and the larger context for farm advisors' and growers' activities, exploring how these groups interact and structure each other's work. By taking a multimethod approach, my aim is to give an ethnographic account of theoretical categories that, though linked, are not always easy to present in a single case study.[1]

As I began planning the research for this project, I met with the director of UC Cooperative Extension in Monterey County and inquired about spending an extended period of time with the county's farm advisors. The Salinas Valley vegetable industry seemed like an ideal case for studying the relationships between science and industry. As I documented in chapter 1, the vegetable industry is an important part of California's farm economy. In addition, because the advisors in Monterey County are known for their emphasis on applied agricultural research, I assumed they would make an ideal case for studying issues related to scientific intervention and repair. The county director and farm advising staff in Monterey County were very receptive to my inquiry, putting very few restrictions on my access to their work. I moved to Salinas in 1997 to begin fieldwork with the advisors, working with them and their staff on a daily basis. In all, I spent a year in this initial period of research in Salinas, returning again in 1999 and 2003 to conduct follow-up research.

My fieldwork activities varied from day to day, depending on the schedule and activities of the advisors themselves; like campus-based scientists, who must balance research, the perpetual money hunt, and education, the farm advisors were always very busy, and it was a challenge to keep up with their activities. They also spent a lot more time sitting in their offices—making phone calls, typing papers, and preparing presentations—than I had expected, and much of the day-to-day maintenance and monitoring of their research projects was performed by technicians. Thus, I spent a lot of time talking with the technicians, and much of my fieldwork on field trials was time I spent with them. I was also able to photograph many aspects of field trials and use the images as instruments for eliciting responses during interviews.[2] In exchange for their involvement in my research, I lent the advisors and their technical staff help in small ways, working on field trials and other tasks. In addition to work on field trials and research in the advisors' on-site facilities, I attended many meetings with them, including the advisors' biweekly staff meetings, educational meetings for growers organized by the advisors, and farm industry meetings.

A large portion of my time was also spent interviewing current and retired farm advisors, their grower clientele, and other persons affiliated with agriculture in the county, including representatives of other government agencies and grower advocate groups. I used a snowball sampling method to select potential interviewees, starting with recommendations from the advisors and then branching further out into the farming community. The interviews were semistructured; though I came prepared with questions, I let the subjects dictate the length of their responses and allowed them to comment on any topic they wished. Nearly all the interviews were tape-recorded and transcribed later. In some cases, informants asked not to have the interview recorded, their right to which was emphasized in an interview consent form. In a few cases, informants also asked that I shut off the tape recorder during their discussion of issues they deemed especially sensitive. Overall, however, I was surprised at how open and extremely helpful informants were in my research. I had been particularly curious about the reaction I would receive from growers—my expectation was that they would be suspicious of my research and hesitant to discuss their company's research interests openly. In fact, just the opposite was the case—in interviews growers were very open and seemed especially grateful for the chance to express their opinions on the state of agricultural research within Cooperative Extension and the UC more generally.

Many of the current and retired farm advisors were interviewed several times, and every interview session with advisors focused on a set of topics that I planned to discuss at that meeting. The following is a breakdown of my interviews: 33 interviews with 11 current advisors in Monterey, Santa Cruz, and San Benito counties;[3] 10 interviews with 6 retired advisors; 8 interviews with 6 members of the advisors' technical staff; 3 interviews with other UC and USDA researchers; 13 interviews with current and retired growers; 4 interviews with representatives of other local government agencies affiliated with agriculture; 4 interviews with representatives of farm industry "interest groups," including commodity boards; and 2 interviews with representatives of small farm interest groups, for a total of 77 interviews. In addition to the interviews I collected during my time in Monterey County, I also draw from a set of 13 interviews conducted with agricultural scientists on the UC Davis campus in 1994, which included extensive discussion of issues related to extension work and the UC's obligations as a land-grant university system.

My attempts to collect historical documents relevant to Cooperative Extension both California-wide and for Monterey County garnered mixed results. The University of California archives at the Bancroft Library in Berkeley are very limited in their scope, confined mostly to the higher echelons of university administration, especially the president's office. Archival resources at UC Davis were disappointing as well—it appears that very few resources related to Cooperative Extension were saved or consolidated at the UC campuses. One important exception to this trend was several years of reports compiled by B. H. Crocheron in the first two decades of UC Cooperative Extension, copies of which I found at the Natural Resources Library on the UC Berkeley campus; these reports form an important part of my analysis in chapter 2.

I also found the documents saved locally, in the offices of Cooperative Extension in Monterey County, very spotty. Although some materials were useful, it was hard to build even the most basic history of the county's extension work, including the staff, their projects, and their clientele over the years. I was fortunate, however, to piece together smaller parts of the history from diverse sources. One of the best resources was the retired advisors and growers who shared their memories with me in interviews. In addition, the Monterey County Farm Bureau had saved the minutes from meetings of its board of directors for the period from approximately 1935 to the present, and the secretary of the organization was kind enough to allow me access to these records. Another important source of data came from records given by the Spreckels Sugar Company to the Monterey County Department of Parks. These records form the bulk of the data used for chapter 4, along with the La Follette Committee's investigations into labor unionism and its repression by industrialists during the 1930s (U.S. Senate 1940).

Notes

Chapter 1

1. The literatures on both labor and water in California agriculture are very large, so I provide only a selected set of references here. For water, see Reisner's (1993) excellent *Cadillac Desert*. For labor, see McWilliams (1939); Galarza (1964; 1977); Reisler (1976); Friedland, Barton, and Thomas (1981); Daniel (1982); Majka and Majka (1982); Thomas (1985); Mitchell (1996); Wells (1996); Vaught (1999). I discuss the history of California farm labor in more detail in chapter 4.

2. My use of the term *built environment* to encompass agricultural environments borrows from William Cronon's and others' contributions to recent environmental history, where the distinction between built and natural environments is being demolished. See Cronon (1991; 1996); White (1995); Mitchell (1996); Mukerji (1997); Stoll (1998).

3. In California a farmer is almost invariably called a grower, and that is the practice I follow in this book.

4. Friedland, Barton, and Thomas (1981); Stoll (1998); Vaught (1999). The path between massive nineteenth-century wheat farming and today's vegetable production is not a straight one. As Vaught (1999) argues, intensive crop production in California prior to about 1920 was often a smaller-scale affair than much of the production today. Nevertheless, the most important institutional factors that allowed for the growth of industrial agriculture in California date to that period.

5. Other common terms include *minor crops* or *specialty crops*, but neither captures the scale and importance of the niche market crops for California agriculture. When heads of iceberg lettuce can be produced and distributed to all regions of the United States for consumption by just about anyone, terms such as these lose their descriptive power.

6. The term *moral economy* comes from J. C. Scott's (1976) classic account of peasant agriculture in Southeast Asia, where he argues that the uncertainty of farming

compels farmers to use methods that provide the most stable yields under many kinds of conditions. Although it would be a mistake to place too much emphasis on the differences between the peasant mode of production described by Scott and that employed in niche industry production, the emphasis for California's growers is on creating the conditions for a reliably *maximal* crop rather than just stable yields.

7. Weber (1946); Mills (1956); Foucault (1979; 1980); Barnes (1988); J. C. Scott (2001).

8. For STS resources, see Kuhn (1970); Collins (1974; 1975; 1985); Latour and Woolgar (1979); Knorr-Cetina (1981; 1999); Bijker, Hughes, and Pinch (1987); Latour (1987; 1988; 1999); Lynch (1985; 1993); Shapin and Schaffer (1985); Traweek (1988); Pickering (1992). For the sociology of agriculture, see Friedland (1974); Friedland and Barton (1975); Friedland, Barton, and Thomas (1981); Busch and Lacy (1983); Buttel, Larson, and Gillespie (1990); Friedland et al. (1991); Rudy et al. (2007). Work in both fields that has explicitly connected these levels of analysis includes Barnes (1977); Gieryn (1983; 1995; 1999); Mukerji (1989; 1997); Fitzgerald (1990; 2003); Busch, Lacy, and Burkhardt (1991); Jasanoff (1995; 2005); Epstein (1996; 2007); Tesh (2000); B. L. Allen (2003); Kleinman (2003); Frickel (2004); Carroll (2006); Frickel and Moore (2006); Warner (2007).

9. My main influence in the use of this metaphor comes from the work of Susan Leigh Star, who has developed the use of the ecology metaphor for understanding scientific practice (Star and Griesemer 1989; Star 1995). Star's use of the ecology metaphor derives from the field of actor-network theory, developed primarily by Michel Callon (1986), John Law (1987; 1994; 2002), and Bruno Latour (1987; 1988; 1990; 1992; 1993; 1995). Although highly influential within the field of science and technology studies, actor-network theory has also been widely criticized for the way that it treats the process of interaction between elements within a network, especially between so-called human and nonhuman elements (Shapin 1988a; Amsterdamska 1990; Schaffer 1991; Collins and Yearley 1992; Jones 1996). For my purposes here, the most important of these criticisms concerns actor-network theory's conception of power, where power is only meaningful in terms of the size and strength of a network, and individual elements in the network are more or less equal in terms of their power (Amsterdamska 1990, 501). In this respect, actor-network theory shows its debt to the work of Foucault, whose work downplayed the role of subjectivity in power relations. In contrast, Star and Griesemer (1989, 390–391) argue for the ecology metaphor as an elaboration of actor-network theory, an alternative that allows for a richer conception of interaction between elements within a network.

10. For STS resources, see Shapin (1988b); Wynne (1989); Ophir and Shapin (1991); Kuklick and Kohler (1996); Mukerji (1997); Gieryn (1998; 2000; 2002); Kohler (2002); Livingstone (2003). For environmental history, see Cronon (1991; 1996);

White (1995); Stoll (1998). Theorists in the field of situated activity are also exploring this dialectic between the context or setting for activity and the action that plays
out there. See Suchman (1987; 1996; 2000); Lave (1988); Hutchins (1995); Engeström
and Middleton (1996); Orr (1996). I describe the role of place in this model in more
detail in chapter 5.

11. The term *fragile power* comes from Mukerji (1989), who describes the fragile
power of oceanographers. Their power is fragile because it is at the discretion of the
state, but the kind of power I am describing here is more diffuse and does not flow
directly from the state in the same way.

12. General resources on ethnomethodology include Garfinkel (1967; 2002);
Heritage (1984); Lynch (1993). The concept of repair comes from ethnomethodological studies of conversation, known as conversation analysis (CA). Specific references on repair include Sacks, Schegloff, and Jefferson (1974); Schegloff, Jefferson,
and Sacks (1977); Schegloff (1992; 1997); Henke (2000). On negotiated order, see
Maines (1977); Fine (1984); Star and Gerson (1986).

13. The improvisational character of interaction is perhaps best described in the
work of Sudnow (1978).

14. However, the work of Schegloff (1992) often suggests that the importance of
repair "transcends the specific concern with intersubjectivity" (1341).

15. Foucault (1972); Geertz (1973; 1983); Hannigan (2006, chs. 3, 4).

16. Habermas (1975); O'Connor (1984; 1998); Rudy (2003). Overall, this process of
repair is similar to Law's (1987) conception of "heterogeneous engineering," where
actors unproblematically mix social and material forms of network building (see
also MacKenzie 1990 and Bowker 1994). Whereas Law's emphasis (and the emphasis
of much empirical work in actor-network approaches) is on network building, my
focus is on network maintenance and adaptation.

17. As one example of many quotations from Jefferson on this subject, in a 1787
letter to George Washington, he wrote, "The wealth acquired by speculation and
plunder, is fugacious in its nature, and fills society with the spirit of gambling. The
moderate and sure income of husbandry begets permanent improvement, quiet life
and orderly conduct, both public and private." University of Virginia, Alderman
Library. Thomas Jefferson to George Washington (1787). See also Danbom (1979);
Daniel (1982, 15).

18. Foucault (1979); Bauman (1991); Law (1994).

19. In this respect, my project follows from activity-centered approaches to understanding state-expert interactions developed by STS scholars such as Jasanoff (1995;
2005); Hilgartner (2000); and Carroll (2006). The multiple and competing interests
surrounding issues of state regulation of industrial production are also emphasized

in recent studies of expertise and the environmental justice movement: Kroll-Smith and Floyd (1997); Fischer (2000); Tesh (2000); B. L. Allen (2003).

Chapter 2

1. McConnell (1953, ch. 1); Danbom (1979); Daniel (1982, 15).

2. McConnell (1953, ch. 1); Mooney and Majka (1995, ch. 2); Sanders (1999).

3. Rosenberg (1976; 1977); Danbom (1979); Marcus (1985).

4. I treat this issue in more detail in chapter 5, where I describe the interaction between farm advisors and growers with respect to field trials, experiments that are often based on the growers' land and intended to demonstrate the advantages of a new way of farming.

5. Scheuring (1995, 42). In fact, Hilgard actually opposed the founding of a university farm in Davisville, years later to become UC Davis (Scheuring 1995, 68).

6. Mowry (1951); Olin (1968); Starr (1985); Danbom (1987); Deverell and Sitton (1994).

7. Olin (1968); Starr (1985, ch. 7); Danbom (1987); Deverell and Sitton (1994).

8. In fact, the first farm advising work in the United States began in the South, under the guidance of Seaman Knapp, an agricultural educator and promoter of the demonstration method of extension, circa 1902 (R. V. Scott 1970, ch. 8).

9. Invariably, farm advising in these early years was a male-dominated profession. As home demonstration advisors were incorporated into the extension system, these advisors were always women.

10. Members of the Country Life movement frequently criticized rural areas for being devoid of entertainment and leisure activities that could provide farm families contact with their neighbors and a chance to get off the farm for at least an evening. Without these ways of cultivating the mind and community, Country Life advocates warned, the best and brightest rural youths would continue to abandon farm life for urban areas.

11. McConnell (1953); Danbom (1979, ch. 3); Mooney and Majka (1995, ch. 3); Scheuring (1995, 81).

12. County Chambers of Commerce were often also active in getting a farm advisor for their county, further allying Cooperative Extension with middle- and upper-class business interests on the local level.

13. Kile (1948). To this day, the AFBF remains among the most powerful farm lobbies in the United States.

14. McWilliams (1939, ch. 2); Beck and Haase (1974, 24); Liebman (1983, 9).

15. McWilliams (1939, ch. 4); Liebman (1983, ch. 1); Almaguer (1994).

Chapter 3

1. All names have been deleted from interview excerpts and replaced with descriptive names. For instance, an advisor who works on issues related to soil and irrigation is identified as Soil/Water, a specialist in plant pathology as PlantPath, an entomologist as Entomol, and so on. For those in the farm industry, I almost always use the generic Grower, although some of them may not actually be in the fields and directly working with the crops. In most cases, though, these industry informants directly influence the production practices of their companies. CRH indicates the author-interviewer.

2. *Salinas Daily Index*, April 9, 1918, 8; April 10, 1918, 4.

3. *Salinas Daily Index*, April 29, 1918, 1.

4. Crocheron, *Progress in Agricultural Extension*, January 1920.

5. Crocheron, *Progress in Agricultural Extension*, June 1927.

6. Crocheron, *Progress in Agricultural Extension*; Anderson (2000).

7. Monterey County Cooperative Extension also included a 4-H advisor and two home demonstration advisors at the time.

8. Although the figure for livestock was much higher than for crops, each cow or sheep requires a large area for grazing. So, with respect to the productive capacities of the land, an acre of land used for crops goes much farther than an acre for livestock.

9. With respect to environmental impacts, the largest growers do not always necessarily represent the area where the most severe problems are to be found. For instance, beginning growers are most likely to be on poor-quality land, often located on steep slopes or with inferior drainage. Farming practices that do not address the unique demands of farming on this land can lead to soil erosion and the pollution of local watersheds (Mountjoy 1996).

10. Draper and Draper (1968); Friedland (1974); Friedland and Barton (1975); Fiske (1979).

11. However, the farm advisors seemed to be universally opposed to the UC's accounting system for tracking advisors' clientele. Each of the current advisors felt that affirmative action was a good idea in principle but that this system was artificial and a waste of their time.

Chapter 4

1. See especially Taylor and Vasey (1936a; 1936b); McWilliams (1939); Galarza (1964; 1977); Reisler (1976); Daniel (1982); Majka and Majka (1982); Thomas (1985); Guerin-Gonzales (1994); Starr (1996); Wells (1996); Vaught (1999).

2. Some works do briefly mention the influence of the UC on farm labor politics in California. McWilliams (1939, 140–141) remarks on the UC economist R. L. Adams's racial typology of California farmworkers and his advice to growers for managing them. Daniel (1982, ch. 2) notes the cozy relationship between early university researchers and the growing farm industry of the 1880s and 1890s. Galarza (1977, 90–93) briefly discusses money contributed to the UC by agribusiness and the production-oriented ethic of agricultural researchers. Scheuring (1995, 142–144) describes a few examples of UC-sponsored activities during the labor unrest of the 1930s that could be characterized as anti-labor.

3. "*Bracero* is the Spanish equivalent of farm hand, meaning one who works with his arms (*brazos*)" (Galarza 1964, 268, n. 2).

4. Reisler (1976, 82). See also Jelinek (1979, 67–68). Vaught (1999) argues that during the early period of niche market farming in California, from about 1880 to 1920, growers had closer connections with farm labor: they would often hire the same group of workers year after year. Niche market farming in the Salinas Valley developed after 1920, when the vegetable industry began shipping railroad cars of produce. By that time niche market growers had developed more formal systems of recruiting farm labor, more in line with the practices I describe in this section.

5. McWilliams (1939); Reisler (1976); Galarza (1977).

6. McWilliams (1939); Galarza (1977); Daniel (1982).

7. These larger, more established unions actually resisted the organization of farmworkers for these very same fears (Daniel 1982, 76–82).

8. Major backers of the Associated Farmers in California included Pacific Gas and Electric, the Southern Pacific Railroad, and the Spreckels Sugar Company. Though grower organizations such as the Agricultural Committee of the State Chamber of Commerce and the California Farm Bureau Federation provided most of the organizational force, these financial and industrial interests provided most of the monetary backing for the Associated Farmers (Daniel 1982, 251–252).

9. Minutes of GSVA member meeting, September 4, 1936, reproduced in U.S. Senate (1940, 27159).

10. In fact, another of the Salinas Valley's most prominent lettuce growers, Bud Antle, was kicked out of the GSVA in 1961, when his company signed a contract with the Teamsters to represent lettuce field-workers (Friedland, Barton, and Thomas

1981, 79). Ironically, in the early 1970s the GSVA membership agreed to Teamster representation as a last-ditch alternative to César Chávez's United Farm Workers.

11. Statement of CASV, released June 25, 1936, reproduced in U.S. Senate (1940, 27127).

12. Smith (1936); McWilliams (1939, 254–259); U.S. Senate (1940, 27192); Starr (1996, 179–189).

13. My discussion in this section mirrors and partly draws upon a similar thread in Scheuring's (1995, 142–145) history of the UC Division of Agriculture.

14. "The Imperial Valley Farm Labor Situation," reproduced in U.S. Senate (1940, 20057).

15. An interesting exception to this conservativism among UC College of Agriculture personnel was expressed by B. H. Crocheron, director of UC Cooperative Extension from 1914 to 1948. In testimony before the La Follette Committee (U.S. Senate 1940, 21741–21745), Crocheron suggested that migrant laborers should be set up on "subsistence homesteads" of two to three acres of land, including a home, which could serve as a base for these families to work at larger farms in the surrounding area. Though this system would have kept the largest farms intact, Crocheron also believed the plan would stabilize the farm workforce, and he emphasized the importance of integrating these newly settled farmworkers into the local community. His testimony, however, casts doubt on the extent to which Crocheron really believed such a system could work. After describing his plan, he notes how difficult it is for a family to survive on a very small farm, expressing his concerns on land subdivision (see chapter 2).

16. See also McWilliams's (1949, 100) discussion of California agriculture, where he explicitly contrasts farming in California with "the American farm tradition."

17. The example of farm advising during the war years parallels work by Mukerji (1989) on oceanography and military funding. Mukerji argues that deep ocean research and the scientists who do this work are funded by government agencies to maintain an emergency reserve of expertise about the ocean, available for deployment in times of war or other crises.

18. Crocheron (1939). A federalist when it came to issues of national policy and planning, Crocheron was not a fan of the New Deal, claiming, "you can't legislate prosperity" (Albaugh 1989, 6).

19. Spreckels (1942b); Rhyner (1942b); Scruggs (1957, ch. 7).

20. Rasmussen (1951, ch. 9); Scruggs (1957, ch. 8); Jelinek (1976, 234–235). In addition to their work with the Mexican National Program, advisors also organized diverse other sources of labor, including schoolchildren on summer vacation, housewives, and German and Italian prisoners of war.

21. Although the respondent called this organization a committee, its actual title was Farm Production Council. I have changed *committee* to *council* in the excerpt for the sake of consistency.

22. In 1890 a federal system of tariffs and duties was enacted to protect and encourage domestic sugar production. These protections, in addition to a federal bounty system paid to sugar producers in many states, set in motion the rapid growth of beet sugar production in the United States. In 1888 there were just two beet-processing factories in the United States (both in California), producing about 2000 tons of beet sugar annually; by 1930, nearly 80 factories were producing over 1 million tons of beet sugar (R. H. Allen 1934, 42).

23. Claus Spreckels's exploits as the "Sugar King" in Hawaii, where he made his fortune planting sugar cane and manufacturing cane sugar, made him infamous before he arrived in the Salinas Valley (Adler 1966).

24. Data on factory operating days are not available for 1899–1904, although R. H. Allen (1934, 45) suggests that the factory ran for 60 days the first three seasons and 90 days subsequently.

25. A Spreckels factory in Watsonville, for instance, was closed immediately after the Salinas facility opened, and the beets normally processed by that factory were shipped to the Salinas plant via rail. Similarly, during the 1920s, years of very low beet production in the Salinas Valley, all the beets produced for a Spreckels plant in Manteca, in the Central Valley, were shipped to the Salinas factory, idling the smaller Manteca facility until the mid-1930s (Pioda, *History*).

26. Beginning in the early 1930s, beet growers even began using field-leveling and bed-shaping techniques developed for vegetable crops. These so-called lettuce beds allowed beet growers to use furrow irrigation for their crop, whereas they previously used a less efficient form, flood irrigation (Pioda, *History*).

27. Along with labor shortages, beet production was hindered by federal programs that limited the acreage of beet production and controlled the price of sugar during the Depression. Depending on the price of sugar and the acreage limitations, beet production could be more or less lucrative, but these decisions were often made too late—growers could only wait for this information for so long before they gave up on a beet contract with Spreckels and planted another crop instead. Therefore, this was an additional limitation on the acreage of beet production early in the war, before restrictions on sugar production were removed for the war effort.

28. In addition to problems with field labor, Spreckels had factory labor problems. Many of the plant workers left for military duty or for work in other industries. As a consequence, Spreckels began placing women workers in beet-processing jobs for the first time (Rhyner 1942a).

29. The *Spreckels Sugar Beet Bulletin* was a grower-oriented journal published six times a year beginning in 1936. It focused on research results and practical advice for growers and was distributed to sugar beet growers throughout the Spreckels territory in California.

30. Purchasing this equipment was not a straightforward matter during the war years because tractors and other forms of farm machinery were rationed. Spreckels had to apply to the U.S. War Production board for the equipment, spare parts, and materials for retrofitting the machinery for beet work (Bullen 1943).

31. Scruggs (1957); Galarza (1964; 1977); Majka and Majka (1982).

32. As an example of the Farm Placement Service's goodwill toward the farm industry in California, Galarza (1977, 44) provides the following labor estimates: "In the Imperial Valley where the total labor force averaged about 15,000 persons—domestics, braceros, and illegals—the declared shortage in 1956 was 15,000. For the entire state in that same year the deficit in manpower for the harvests was estimated at over 111,000, roughly equivalent to all the domestics."

33. Spreckels (1954, 18). The interlocutor in this exchange is likely Austin Armer, the agricultural engineer who headed the Spreckels mechanization campaign.

34. Land prices were also an important factor in this slide, becoming too high to justify the lower commodity prices of sugar beets, dry beans, barley, and other nonvegetable crops that were once popular with growers in Monterey County.

35. Schegloff, Jefferson, and Sacks (1977); Schegloff (1992; 1997).

Chapter 5

1. It is important to emphasize that this point is about the *goal* of most laboratory-based research, where the aim is to transcend place. A large literature in laboratory studies of scientific practice shows, however, that this goal is always only partly fulfilled and that laboratory research is also contingent on the local conditions of practice. See especially Collins (1974); Latour and Woolgar (1979); Lynch (1985); Shapin and Schaffer (1985); Shapin (1988b); Traweek (1988).

2. In addition, much of the literature on science and place tends to draw sharp boundaries between laboratory- and field-based science, whereas applied agricultural science is a blend of techniques often associated with either lab or field science. One exception to this trend is the work of Latour, especially his study of Pasteur's work on a vaccine for anthrax (Latour 1988; see also Latour 1995). Latour details the connections that Pasteur made between laboratory and field, describing the way that the field was remade, or translated, to resemble the lab: "All the Pasteurian 'applications' were 'diffused,' as we say, only if it was previously possible to create *in situ* the conditions of the laboratory" (Latour 1988, 90). But the lab-centeredness

of Latour's analysis is a deficiency here, because translation is not always about making a lab of the field. In fact, extension work in agriculture is often predicated on making a field trial seem just like any other grower's field, in order to make the results found there more convincing to those who work in fields, not labs.

3. Similar results are reported by Wynne (1989).

4. Mukerji (1989); Shapin (1989); Barley and Bechky (1994); Orr (1996); Barley and Orr (1997); Sims (1999).

5. Concerned about issues of accuracy, the research team decided to weigh celery heads from each plot that were graded into a specific category, for example, taking the weight of a 30 from each plot. They were then able to compare the weights of 30s across the plots to check on the consistency of the farmworker. The advisor later told me that they did not need to make this check because the differences between weights of a given grade proved to be very small across plots. Statistically speaking, the differences were so small that this accuracy was not very likely to be produced by chance alone.

6. Of course, if the grower is accustomed to thinking of yield in terms of weight, this would be a perfectly acceptable way to take harvest data. The potato field trial in figure 5.2 is one such example. But even crops where the yield is measured by weight often depend on standardized fieldwork practices to ensure valid comparisons of the experimental and control areas.

7. This point echoes STS work on testing: the results of a test do not always definitively say "it works" (or not). See MacKenzie (1990); Sims (1999).

8. Especially relevant here are those writers who emphasize the embeddedness of representational practices within specific cultures. See Rudwick (1976); Pickering (1984); Lave (1988); Shapin (1994); Henderson (1999); Kohler (2002). Other authors, however, have described how visual and numerical representations can break away from the confines of specific places, making information mobile in unique and very powerful ways (Latour 1986; 1995; Porter 1995). Without the requisite practices for seeing and counting already in place, the universalistic power of these representations may be overstated. If action-at-a-distance really did work in applied agricultural science, then the local character of farming places—and advisors' location in these places—would matter much less for repair.

9. In chapter 6 I examine in more detail advisors' struggles to promote environmental change.

Chapter 6

1. NRC (1989, ch. 1); Ongley (1996, ch. 1); USEPA (2003, ch. 2).

2. Beck (1992); Beck, Giddens, and Lash (1994).

3. Beck (1992, 32); Van Loon (2002); see also Rycroft (1991); Luke (1999); Fischer (2000).

4. Spaargaren and Mol (1992); Mol and Spaargaren (1993); Mol (1996); Hajer (1995); Cohen (1997).

5. USEPA (2003). *Nonpoint* means that agriculture does not spill all its pollutants out at a single source point, such as a pipe spewing wastewater from a factory directly into a river.

6. Although I am using the past tense to describe these events, in 2007 most of the conditions I describe here remain the same.

7. Growers and input manufacturers have long bemoaned the lengthy (and costly) approval process that the EPA uses before registering a chemical for a particular crop. Any time a chemical company develops a new product, it has to undergo a lengthy period of testing to prove its utility and safety *for a specific crop*. For so-called minor crops, such as the vegetables grown in the Salinas Valley, there is often little incentive to go through this process, especially for specialty crops that give the manufacturers a smaller return on their registration costs.

8. My use of this line of questioning was at least partly influenced by Rose's (1994, ch. 5) study of hip-hop artists. Rose found that many women rappers are hesitant to call themselves feminists, even though their raps often voice common feminist concerns about gender politics and relationships. By asking a leading question like this, I ran the risk of biasing the advisors' responses, but I believe that this risk was minimal in this case. First, I had been working with the advisors for more than six months when I conducted these interviews, and they were by then comfortable with me, and seemed to feel no need to impress me. Second, and perhaps more important, advisors spend a lot of time working on environmental problems; they are not making up an environmental identity out of nothing.

9. The most prominent of these problems was the intrusion of seawater into groundwater supplies along the Monterey Bay coast. Although this problem had existed for decades, increased pumping of groundwater for crop irrigation had brought seawater further inland during the late 1990s and early 2000s, until it nearly reached the very valuable farming lands just outside the City of Salinas.

10. Pest Control Advisor (PCA) represents an occupation that parallels Cooperative Extension, but in the private sector. Although nonpublic advisors have been around for at least as long as Cooperative Extension itself, the growth of the PCA profession occurred in California out of legislative (and market) necessity. California law requires all farmers to receive a PCA's recommendation for any application of pesticides (including insecticides, herbicides, and fungicides) on food crops. The intent of the law is to make sure expert advice is given to growers on issues of pesticide application, ensuring that only pesticides that are registered for certain crops by the

EPA are applied to those crops. A PCA's advice should also make for sensible and judicious use of chemical applications; however, many PCAs are employed by commercial agricultural chemical dealers and applicators. Therefore, many of those responsible for providing advice on the application of pesticides are also salespeople.

11. With a program method of pesticide application, pests are controlled by systematic pesticide spray cycles. Therefore, the pests themselves are growing and evolving with these chemicals as a constant part of their environment. Given that the life of a disease pathogen or an insect is very short, it does not take long for newer generations to develop a resistance to the chemicals. Intensifying the program to include more sprayings and at higher doses just exacerbates the problem; the pests will eventually develop an even stronger resistance until the chemical is useless against them at any concentration. In many cases, pests begin to show signs of resistance to a new chemical within a few years of its introduction.

12. There are products such as so-called soft pesticides that are an attempt to maintain use of chemical inputs while mitigating the effects of "harder" chemicals.

Chapter 7

1. Gilbert and Akor (1988); Harper (2001); Gloy, Hyde, and LaDue (2002); CDFA (2006); NYSDAM (2006).

2. Just a small sampling of this literature includes Schlosser (2001); Nestle (2002); and Pollan (2006).

3. Callon (1998); McCloskey (1998); B. L. Allen (2003, ch. 3); MacKenzie (2006).

4. For example, in other recent work I have examined the history of land use and flood control in New Orleans, particularly in light of the devastation caused by Hurricane Katrina in 2005 (Henke 2007). New Orleans is a built environment that reflects the interests of those who had a key role in shaping the city's growth into a major port and site for industry along the Mississippi River. Flood control for the city focused heavily on the use of levees because this technology was promoted by both U.S. Army Corps of Engineers experts and wealthy landowners who wanted to preserve as much land as possible for development. In a recurring series of flood crises, the emphasis has always been on preserving the levee system rather than implementing more radical changes to the flood control system, even leading to the deliberate flooding of St. Bernard Parish in 1927. Levees downriver of New Orleans were dynamited in order to relieve increasing water pressure on the city's levees, thus preserving New Orleans at the expense of the thousands of residents who were flooded out of St. Bernard Parish. While Hurricane Katrina created far more damage, the point is that examining the ecology of material infrastructure, interests, and ideologies places Katrina in context—the events become part of a long-term series

of choices and struggles over the shape of the ecology of power in New Orleans. Furthermore, actors' responses to repair crises such as these reveal their attempts to either maintain or transform this structure.

5. Despite the importance of PCAs as a source of private information and advice, UC farm advisors remain the main source for the PCAs themselves to gain new knowledge about farming practices and technologies. PCAs need to gain continuing education credits each year to maintain their PCA license, and farm advisors' meetings and educational seminars constitute the main source of this learning. Advisor bulletins, reports, and field trial demonstration events are other sources for UC advisors to transfer information to PCAs.

Appendix

1. In this way, I am following the extended case method approach proposed by Michael Burawoy (1991) but also using a vision of ethnography that is not synonymous with participant observation.

2. For a good example of the photo interview method, see Harper (1987).

3. Santa Cruz and San Benito counties border Monterey County, and many of the advisors assigned to these counties have cross-county assignments to work in adjacent counties. Therefore, I interviewed some of the advisors working in these other counties as well.

References

Adams, R. L. 1921. *Farm Management: A Text-Book for Student, Investigator, and Investor.* New York: McGraw Hill.

Adams, R. L., and T. R. Kelly. 1918. A Study of Farm Labor in California. Circular 193. University of California Experiment Station, Berkeley.

Adler, J. 1966. *Claus Spreckels: The Sugar King in Hawaii.* Honolulu: University of Hawaii Press.

Albaugh, R. 1989. The Golden Age of Extension: Backtrailing During the Barehanded Days. Speech presented May 19.

Allen, B. L. 2003. *Uneasy Alchemy: Citizens and Experts in Louisiana's Chemical Corridor Disputes.* Cambridge, Mass.: MIT Press.

Allen, R. H. 1934. Economic History of Agriculture in Monterey County, California, During the American Period. Ph.D. diss., Department of Agricultural Economics, University of California, Berkeley.

Almaguer, T. 1994. *Racial Fault Lines: The Historical Origins of White Supremacy in California.* Berkeley: University of California Press.

Amsterdamska, O. 1990. Surely You Are Joking, Monsieur Latour! *Science, Technology, and Human Values* 15: 495–504.

Anderson, B. 2000. *The Salinas Valley: A History of America's Salad Bowl.* Salinas, Calif.: Monterey County Historical Society.

Armer, A. 1944. Plan Planting for Mechanical Harvest. *Spreckels Sugar Beet Bulletin* 8: 1–2.

———. 1947. A Review of the 1946 Mechanical Harvest. *Spreckels Sugar Beet Bulletin* 11: 14.

———. 1951. Prepare for a Shortage of Field Labor. *Spreckels Sugar Beet Bulletin* 15: 10–11.

Athanasiou, T. 1996. The Age of Greenwashing. *Capitalism, Nature, Socialism* 7: 1–36.

Auerbach, J. S. 1966. *Labor and Liberty: The La Follette Committee and the New Deal.* Indianapolis: Bobbs-Merrill.

Austin, A. 2002. Advancing Accumulation and Managing Its Discontents: The U.S. Antienvironmental Countermovement. *Sociological Spectrum* 22: 71–105.

Barley, S. R., and B. A. Bechky. 1994. In the Backrooms of Science: The Work of Technicians in Science Labs. *Work and Occupations* 21: 85–126.

Barley, S. R., and J. E. Orr. 1997. Introduction: The Neglected Workforce. In *Between Craft and Science: Technical Work in U.S. Settings*, ed. S. R. Barley and J. E. Orr. Ithaca, N.Y.: Cornell University Press.

Barnes, B. 1977. *Interests and the Growth of Knowledge.* London: Routledge and Kegan Paul.

———. 1988. *The Nature of Power.* Urbana: University of Illinois Press.

Bauman, Z. 1991. *Modernity and Ambivalence.* Cambridge: Polity Press.

Beck, U. 1992. *Risk Society: Towards a New Modernity.* London: Sage.

Beck, U., A. Giddens, and S. Lash. 1994. *Reflexive Modernization: Politics, Tradition, and Aesthetics in the Modern Social Order.* Stanford, Calif.: Stanford University Press.

Beck, W. A., and Y. D. Haase. 1974. *Historical Atlas of California.* Norman: University of Oklahoma Press.

Bijker, W. E., T. P. Hughes, and T. J. Pinch, eds. 1987. *The Social Construction of Technological Systems: New Directions in the Sociology and History of Technology.* Cambridge, Mass.: MIT Press.

Bourdieu, P. 1990. *The Logic of Practice.* New York: Cambridge University Press.

———. 2004. *Science of Science and Reflexivity.* Chicago: University of Chicago Press.

Bowers, W. L. 1974. *The Country Life Movement in America, 1900–1920.* Port Washington, N.Y.: Kennikat Press.

Bowker, G. C. 1994. *Science on the Run: Information Management and Industrial Geophysics at Schlumberger, 1920–1940.* Cambridge, Mass.: MIT Press.

Bullen, A. C. 1943. Beet Harvesting Machinery for 1943. *Spreckels Sugar Beet Bulletin* 7: 11–12.

Burawoy, M. 1991. The Extended Case Method. In *Ethnography Unbound: Power and Resistance in the Modern Metropolis*, by M. Burawoy et al., 271–290. Berkeley: University of California Press.

Busch, L., and W. B. Lacy. 1983. *Science, Agriculture, and the Politics of Research.* Boulder, Colo.: Westview Press.

Busch, L., W. B. Lacy, J. Burkhardt. 1991. *Plants, Power, and Profit: Social, Economic, and Ethical Consequences of the New Biotechnologies.* Cambridge, Mass.: Blackwell.

Buttel, F. H., O. F. Larson, and G. W. Gillespie Jr. 1990. *The Sociology of Agriculture.* Westport, Conn.: Greenwood Press.

Callon, M. 1986. Some Elements of a Sociology of Translation: Domestication of the Scallops and the Fishermen of St. Brieuc Bay. In *Power, Action, and Belief: A New Sociology of Knowledge?* ed. J. Law, 196–233. London: Routledge and Kegan Paul.

———, ed. 1998. *The Laws of the Markets.* Oxford: Blackwell.

Carroll, P. 2006. *Science, Culture, and Modern State Formation.* Berkeley: University of California Press.

Carson, Rachel. 1962. *Silent Spring.* Boston: Houghton Mifflin.

CDFA (California Department of Food and Agriculture). 2006. California Dairy Statistics and Trends 2006. Sacramento.

Chambers, R., A. Pacey, and L. A. Thrupp. 1989. *Farmer First: Farmer Innovation and Agricultural Research.* Intermediate Technology Publications. Warwickshire, U.K.: Practical Action Publishing.

CLRB (California Lettuce Research Board). 2007. Activities, Issues, and Accomplishments: An Evaluation of the California Lettuce Research Board. <http://www.calettuceresearchboard.org/>.

Cochrane, W. W. 1993. *The Development of American Agriculture: A Historical Analysis.* 2d ed. Minneapolis: University of Minnesota Press.

Cohen, M. J. 1997. Risk Society and Ecological Modernisation: Alternative Visions for Post-Industrial Nations. *Futures* 29: 105–119.

Collins, H. M. 1974. The TEA Set: Tacit Knowledge and Scientific Networks. *Science Studies* 4: 165–186.

———. 1975. The 7 Sexes: A Study in the Sociology of a Phenomenon, or the Replication of Experiments in Physics. *Sociology* 9: 205–224.

———. 1985. *Changing Order: Replication and Induction in Scientific Practice.* Beverly Hills, Calif.: Sage.

Collins, H. M., and S. Yearley. 1992. Epistemological Chicken. In *Science as Practice and Culture*, ed. A. Pickering, 301–326. Chicago: University of Chicago Press.

Crane, C. E. 1943. Blocking 6 Rows at Once: A. L. Roddick and Sons Successfully Use Dixie Blocker on Bed Planting. *Spreckels Sugar Beet Bulletin* 7: 17, 19.

Crocheron, B. H. 1914. The County Farm Adviser. Circular No. 112. University of California College of Agriculture, Berkeley.

———. 1918. Summary of the Annual Reports of the Farm Advisors of California. University of California Agricultural Extension Service, Berkeley.

———. 1919–1935. *Progress in Agricultural Extension.* Series of reports. University of California Agricultural Extension Service, Berkeley.

———. 1939. After Twenty-five Years: An Appraisal. Address given at University of California Agricultural Experiment Station conference, Berkeley, January 7.

———. 1946. The Years Between: Being an Account of Some of the Activities of the Agricultural Extension Service of the University of California During the Second World War. Address given at University of California Agricultural Experiment Station conference, Berkeley, January 7.

Cronon, W. 1991. *Nature's Metropolis: Chicago and the Great West.* New York: W.W. Norton.

———, ed. 1996. *Uncommon Ground: Rethinking the Human Place in Nature.* New York: W.W. Norton.

Danbom, D. B. 1979. *The Resisted Revolution: Urban America and the Industrialization of Agriculture, 1900–1930.* Ames: Iowa State University Press.

———. 1987. *The World of Hope: Progressives and the Struggle for an Ethical Public Life.* Philadelphia: Temple University Press.

———. 1995. *Born in the Country: A History of Rural America.* Baltimore: Johns Hopkins University Press.

Daniel, C. E. 1982. *Bitter Harvest: A History of California Farmworkers, 1870–1941.* Berkeley: University of California Press.

Deverell, W., and T. Sitton, eds. 1994. *California Progressivism Revisited.* Berkeley: University of California Press.

Diamond, J. 2005. *Collapse: How Societies Choose to Fail or Succeed.* New York: Viking.

Dimitri, C., A. Effland, and N. Conklin. 2005. The 20th Century Transformation of U.S. Agriculture and Farm Policy. Economic Information Bulletin No 3. Washington, D.C.: USDA Economic Research Service.

Draper, A., and H. Draper. 1968. *The Dirt on California: Agribusiness and the University.* Berkeley: Independent Socialist Clubs of America.

Easterling, W. E., et al. 2007. Food, Fibre and Forest Products. Climate Change 2007: Impacts, Adaptation and Vulnerability. In *Fourth Assessment Report of the*

Intergovernmental Panel on Climate Change, ed. M. L. Parry et al., 273–313. Cambridge: Cambridge University Press.

Engeström, Y., and D. Middleton, eds. 1996. *Cognition and Communication at Work.* New York: Cambridge University Press.

Epstein, S. 1996. *Impure Science: Aids, Activism, and the Politics of Science.* Berkeley: University of California Press.

———. 2007. *Inclusion: The Politics of Difference in Medical Research.* Chicago: University of Chicago Press.

Fine, G. A. 1984. Negotiated Orders and Organizational Cultures. *Annual Review of Sociology* 10: 239–262.

Fischer, F. 2000. *Citizens, Experts, and the Environment: The Politics of Local Knowledge.* Durham: Duke University Press.

Fiske, E. P. 1979. The College and Its Constituency: Rural and Community Development at the University of California, 1875–1978. Ph.D. diss., University of California, Davis.

Fitzgerald, D. 1990. *The Business of Breeding: Hybrid Corn in Illinois, 1890–1940.* Ithaca, N.Y.: Cornell University Press.

———. 2003. *Every Farm a Factory: The Industrial Ideal in American Agriculture.* New Haven: Yale University Press.

Foucault, M. 1972. *The Archaeology of Knowledge.* New York: Pantheon Books.

———. 1979. *Discipline and Punish: The Birth of the Prison.* New York: Random House.

———. 1980. *Power/Knowledge: Selected Interviews and Other Writings, 1972–1977.* New York: Pantheon Books.

Frickel, S. 2004. *Chemical Consequences: Environmental Mutagens, Scientist Activism, and the Rise of Genetic Toxicology.* New Brunswick, N.J.: Rutgers University Press.

Frickel, S., and K. Moore, eds. 2006. *The New Political Sociology of Science: Institutions, Networks, and Power.* Madison: University of Wisconsin Press.

Friedland, W. H. 1974. Social Sleepwalkers: Scientific and Technological Research in California Agriculture. Research Monograph 13. University of California, Davis, Department of Applied Behavioral Sciences.

Friedland, W. H., and A. Barton. 1975. Destalking the Wily Tomato: A Case Study in the Social Consequences in California Agricultural Research. Research Monograph 15. University of California, Davis, Department of Applied Behavioral Sciences.

Friedland, W. H., A. Barton, and R. J. Thomas. 1981. *Manufacturing Green Gold: Capital, Labor, and Technology in the Lettuce Industry*. New York: Cambridge University Press.

Friedland, W. H., L. Busch, F. H. Buttel, and A. P. Rudy, eds. 1991. *Towards a New Political Economy of Agriculture*. Boulder, Colo.: Westview Press.

Galarza, E. 1964. *Merchants of Labor: The Mexican Bracero Story*. Charlotte, N.C.: McNally and Loftin.

———. 1977. *Farm Workers and Agri-Business in California, 1947–1960*. South Bend, Ind.: University of Notre Dame Press.

Garfinkel, H. 1967. *Studies in Ethnomethodology*. Englewood Cliffs, N.J.: Prentice Hall.

———. 2002. *Ethnomethodology's Program: Working out Durkheim's Aphorism*. Lanham, Md.: Rowman and Littlefield.

Geertz, C. 1973. *The Interpretation of Cultures*. New York: HarperCollins.

———. 1983. *Local Knowledge: Further Essays in Interpretive Anthropology*. New York: Basic Books.

Gieryn, T. F. 1983. Boundary-Work and the Demarcation of Science from Non-Science: Strains and Interests in Professional Ideologies of Science. *American Sociological Review* 48: 781–795.

———. 1995. Boundaries of Science. In *Handbook of Science and Technology Studies*, ed. S. Jasanoff et al., 393–443. Thousand Oaks, Calif.: Sage.

———. 1998. Biotechnology's Private Parts (and Some Public Ones). In *Private Science: Biotechnology and the Rise of the Molecular Sciences*, ed. A. Thackray, 219–253. Philadelphia: University of Pennsylvania Press.

———. 1999. *Cultural Boundaries of Science: Credibility on the Line*. Chicago: University of Chicago Press.

———. 2000. A Space for Place in Sociology. *Annual Review of Sociology* 26: 463–496.

———. 2002. Three Truth-Spots. *Journal of the History of the Behavioral Sciences* 38: 113–132.

Gilbert, J., and R. Akor. 1988. Increasing Structural Divergence in U.S. Dairying: California and Wisconsin since 1950. *Rural Sociology* 53: 56–72.

Gloy, B. A., J. Hyde, and E. L. LaDue. 2002. Dairy Farm Management and Long-Term Farm Financial Performance. *Agricultural and Resource Economics Review* 31: 233–247.

Guerin-Gonzales, C. 1994. *Mexican Workers and American Dreams: Immigration, Repatriation, and California Farm Labor, 1900–1939*. New Brunswick, N.J.: Rutgers University Press.

Habermas, J. 1975. *Legitimation Crisis*. Boston: Beacon Press.

Hajer, M. A. 1995. *The Politics of Environmental Discourse: Ecological Modernization and the Policy Process*. New York: Oxford University Press.

Hallberg, M. C. 2001. *Economic Trends in U.S. Agriculture and Food Systems since World War II*. Ames: Iowa State University Press.

Hannigan, J. 2006. *Environmental Sociology: A Social Constructionist Perspective*. 2d ed. New York: Routledge.

Harper, D. 1987. *Working Knowledge: Skill and Community in a Small Shop*. Berkeley: University of California Press.

———. 2001. *Changing Works: Visions of a Lost Agriculture*. Chicago: University of Chicago Press.

Henderson, K. 1999. *On Line and on Paper: Visual Representations, Visual Culture, and Computer Graphics in Design Engineering*. Cambridge, Mass.: MIT Press.

Henke, C. R. 2000. The Mechanics of Workplace Order: Toward a Sociology of Repair. *Berkeley Journal of Sociology* 44: 55–81.

———. 2007. Situation Normal? Repairing a Risky Ecology. *Social Studies of Science* 37: 135–142.

Heritage, J. 1984. *Garfinkel and Ethnomethodology*. Cambridge: Polity Press.

Hightower, J. 1973. *Hard Tomatoes, Hard Times: A Report of the Agribusiness Accountability Project on the Failure of America's Land Grant College Complex*. Cambridge, Mass.: Schenckman.

Hilgartner, S. 2000. *Science on Stage: Expert Advice as Public Drama*. Stanford, Calif.: Stanford University Press.

Hutchins, E. 1995. *Cognition in the Wild*. Cambridge, Mass.: MIT Press.

Jasanoff, S. 1995. *Science at the Bar: Law, Science, and Technology in America*. Cambridge, Mass.: Harvard University Press.

———. 2005. *Designs on Nature: Science and Democracy in Europe and the United States*. Princeton, N.J.: Princeton University Press.

Jelinek, L. J. 1976. The California Farm Bureau Federation, 1919–1964. Ph.D. diss., Department of History, University of California, Los Angeles.

——. 1979. *Harvest Empire: A History of California Agriculture*. San Francisco: Boyd and Fraser.

Jones, M. P. 1996. Posthuman Agency: Between Theoretical Traditions. *Sociological Theory* 14: 290–310.

Kile, O. M. 1948. *The Farm Bureau through Three Decades*. Baltimore: Waverly Press.

Kleinman, D. 2003. *Impure Cultures: University Biology and the World of Commerce*. Madison: University of Wisconsin Press.

Knorr-Cetina, K. 1981. *The Manufacture of Knowledge: An Essay on the Constructivist and Contextual Nature of Science*. Oxford: Pergamon Press.

——. 1999. *Epistemic Cultures: How the Sciences Make Knowledge*. Cambridge, Mass.: Harvard University Press.

Kohler, R. E. 2002. *Landscapes and Labscapes: Exploring the Lab-Field Border in Biology*. Chicago: University of Chicago Press.

Kroll-Smith, S., and H. H. Floyd. 1997. *Bodies in Protest: Environmental Illness and the Struggle over Medical Knowledge*. New York: New York University Press.

Kuhn, T. S. 1970. *The Structure of Scientific Revolutions*. 2d ed. Chicago: University of Chicago Press.

Kuklick, H., and R. E. Kohler, eds. 1996. *Osiris Volume 11: Science in the Field*. Chicago: University of Chicago Press.

Lambdin, R. 1943. Costs Cut with Dixie Thinner. *Spreckels Sugar Beet Bulletin* 7: 29.

Latour, B. 1986. Visualization and Cognition: Thinking with Eyes and Hands. In *Knowledge and Society: Studies in the Sociology of Culture Past and Present*, ed. H. Kuklick and E. Long, 1–40. Greenwich, Conn.: JAI Press.

——. 1987. *Science in Action: How to Follow Scientists and Engineers through Society*. Cambridge, Mass.: Harvard University Press.

——. 1988. *The Pasteurization of France*. Cambridge, Mass.: Harvard University Press.

——. 1990. Postmodern? No, Simply Amodern! Steps Towards an Anthropology of Science. *Studies in the History and Philosophy of Science* 21: 145–171.

——. 1992. Where Are the Missing Masses? The Sociology of a Few Mundane Artifacts. In *Shaping Technology/Building Society*, ed. W. Bijker and J. Law, 225–258. Cambridge, Mass.: MIT Press.

——. 1993. *We Have Never Been Modern*. Cambridge, Mass.: Harvard University Press.

——. 1995. The "Pédofil" of Boa Vista: A Photo-Philosophical Montage. *Common Knowledge* 4: 145–187.

——. 1999. *Pandora's Hope: Essays on the Reality of Science Studies.* Cambridge, Mass.: Harvard University Press.

Latour, B., and S. Woolgar. 1979. *Laboratory Life: The Construction of Scientific Facts.* Princeton, N.J.: Princeton University Press.

Lave, J. 1988. *Cognition in Practice: Mind, Mathematics and Culture in Everyday Life.* New York: Cambridge University Press.

Law, J. 1987. Technology and Heterogeneous Engineering: The Case of Portuguese Expansion. In *The Social Construction of Technological Systems: New Directions in the Sociology and History of Technology,* ed. W. E. Bijker, T. P. Hughes, and T. J. Pinch, 111–134. Cambridge, Mass.: MIT Press.

——. 1994. *Organizing Modernity.* Cambridge, Mass.: Blackwell.

——. 2002. *Aircraft Stories: Decentering the Object in Technoscience.* Durham: Duke University Press.

Liebman, E. 1983. *California Farmland: A History of Large Agricultural Landholdings.* Montclair, N.J.: Rowman and Allanheld.

Liss, S. 1953. Farm Wage Boards under the Cooperative Extension Service During World War II. *Agricultural History* 27: 103–108.

Livingstone, D. N. 2003. *Putting Science in Its Place: Geographies of Scientific Knowledge.* Chicago: University of Chicago Press.

Luke, T. 1999. Eco-Managerialism: Environmental Studies as a Power/Knowledge Formation. In *Living with Nature: Environmental Politics as Cultural Discourse,* ed. F. Fischer and M. A. Hajer, 103–120. New York: Oxford University Press.

Lynch, M. 1985. *Art and Artifact in Laboratory Science: A Study of Shop Work and Shop Talk in a Research Laboratory.* Boston: Routledge.

——. 1993. *Scientific Practice and Ordinary Action: Ethnomethodology and Social Studies of Science.* New York: Cambridge University Press.

Lyson, T. A. 2004. *Civic Agriculture: Reconnecting Farm, Food, and Community.* Medford, Mass.: Tufts University Press.

MacKenzie, D. 1990. *Inventing Accuracy: A Historical Sociology of Nuclear Missile Guidance.* Cambridge, Mass.: MIT Press.

——. 2006. *An Engine, Not a Camera: How Financial Models Shape Markets.* Cambridge, Mass.: MIT Press.

Maines, D. R. 1977. Social Organization and Social Structure in Symbolic Interactionist Thought. *Annual Review of Sociology* 3: 235–259.

Majka, L. C., and T. J. Majka. 1982. *Farmworkers, Agribusiness, and the State*. Philadelphia: Temple University Press.

Marcus, A. I. 1985. *Agricultural Science and the Quest for Legitimacy: Farmers, Agricultural Colleges, and Experiment Stations, 1870–1890*. Ames: Iowa State University Press.

McCabe, M. 1998a. Nitrate-Laced Water Sickens Town; Monterey County Warns Residents Not to Use Taps. *San Francisco Chronicle*, May 12, A1.

———. 1998b. Monterey County OKs Well for Its Poorest Town. *San Francisco Chronicle*, May 26, A11.

McCloskey, D. N. 1998. *The Rhetoric of Economics*. 2d ed. Madison: University of Wisconsin Press.

McConnell, G. 1953. *The Decline of Agrarian Democracy*. Berkeley: University of California Press.

MCFB (Monterey County Farm Bureau). 1943. Minutes of Directors' Meeting, July 9.

———. 1945. Minutes of Directors' Meeting, September 13.

———. 1946. Minutes of Directors' Meeting, March 13.

McWilliams, C. 1939. *Factories in the Field: The Story of Migratory Farm Labor in California*. Boston: Little, Brown.

———. 1949. *California: The Great Exception*. Westport, Conn.: Greenwood Press.

MCWRA (Monterey County Water Resources Agency). 2002. Nitrate Management Survey Results Report, 2001. Salinas, Calif.

Mills, C. W. 1956. *The Power Elite*. New York: Oxford University Press.

Mitchell, D. 1996. *The Lie of the Land: Migrant Workers and the California Landscape*. Minneapolis: University of Minnesota Press.

Mol, A. P. J. 1996. Ecological Modernisation and Institutional Reflexivity: Environmental Reform in the Late Modern Age. *Environmental Politics* 5: 302–323.

Mol, A. P. J., and G. Spaargaren. 1993. Environment, Modernity, and the Risk Society: The Apocalyptic Horizon of Environmental Reform. *International Sociology* 8: 431–459.

Mooney, P. H., and T. J. Majka. 1995. *Farmers' and Farmworkers' Movements: Social Protest in American Agriculture*. New York: Twayne Publishers.

Mountjoy, D. C. 1996. Ethnic Diversity and the Patterned Adoption of Soil Conservation in the Strawberry Hills of Monterey, California. *Society and Natural Resources* 9: 339–357.

Mowry, G. E. 1951. *The California Progressives*. Berkeley: University of California Press.

Mukerji, C. 1989. *A Fragile Power: Scientists and the State*. Princeton, N.J.: Princeton University Press.

———. 1997. *Territorial Ambitions and the Gardens of Versailles*. New York: Cambridge University Press.

Nestle, M. 2002. *Food Politics: How the Food Industry Influences Nutrition and Health*. Berkeley: University of California Press.

Noble, D. F. 1977. *America by Design: Science, Technology, and the Rise of Corporate Capitalism*. New York: Oxford University Press.

———. 1984. *Forces of Production: A Social History of Industrial Automation*. New York: Random House.

Norris, F. 1901. *The Octopus: A Story of California*. New York: Penguin.

NRC (National Research Council). 1989. *Alternative Agriculture*. Washington, D.C.: National Academies Press.

NYSDAM (New York State Department of Agriculture and Markets). 2006. New York State Dairy Statistics, Annual Summary. Division of Milk Control and Dairy Services. Albany.

O'Connor, J. 1984. *Accumulation Crisis*. New York: Blackwell.

———. 1998. *Natural Causes: Essays in Ecological Marxism*. New York: Guilford Press.

Olin, S. C., Jr. 1968. *California's Prodigal Sons: Hiram Johnson and the Progressives, 1911–1917*. Berkeley: University of California Press.

Ongley, E. D. 1996. Control of Water Pollution from Agriculture. FAO Irrigation and Drainage Paper 55. Rome: UN Food and Agriculture Organization.

Ophir, A., and S. Shapin. 1991. The Place of Knowledge: A Methodological Survey. *Science in Context* 4: 3–21.

Orr, J. E. 1996. *Talking About Machines: An Ethnography of a Modern Job*. Ithaca, N.Y.: Cornell University Press.

Peterson, K. 1981. Spreckels Plant to Close Doors. *Monterey Peninsula Herald*, 1.

Pickering, A. 1980. The Role of Interests in High-Energy Physics: The Choice between Charm and Colour. In *The Social Process of Scientific Investigation*, ed. K. D. Knorr, R. Krohn, and R. P. Whitley, 107–138. Dordrecht: D. Reidel.

———. 1984. Against Putting the Phenomena First: The Discovery of the Weak Neutral Current. *Studies in History and Philosophy of Science* 15: 85–117.

———, ed. 1992. *Science as Practice and Culture*. Chicago: University of Chicago Press.

Pioda, C. L. 1943. Mexican Nationals Salvage Beet Crop. *Spreckels Honey Dew News*, August, 14.

———. n.d. *Chronological History, Spreckels Sugar Company, 1897 through 1945*. Salinas, Calif.: Monterey County Parks Department.

Pollan, M. 2006. *The Omnivore's Dilemma: A Natural History of Four Meals*. New York: Penguin.

Porter, T. 1995. *Trust in Numbers: The Pursuit of Objectivity in Science and Public Life*. Princeton, N.J.: Princeton University Press.

Rasmussen, W. D. 1951. *A History of the Emergency Farm Labor Supply Program, 1943–1947*. Washington, D.C.: USDA, Bureau of Agricultural Economics.

Reisler, M. 1976. *By the Sweat of Their Brow: Mexican Immigrant Labor in the United States, 1900–1940*. Westport, Conn.: Greenwood Press.

Reisner, M. 1993. *Cadillac Desert: The American West and Its Disappearing Water*. Rev. ed. New York: Penguin.

Rhyner, R. 1942a. The Pan Sheet. *Spreckels Honey Dew News*, September, 16–17.

———. 1942b. The Pan Sheet. *Spreckels Honey Dew News*, October, 9.

Rogers, E. M. 1958. Categorizing the Adopters of Agricultural Practices. *Rural Sociology* 23: 345–354.

———. 1983. *Diffusion of Innovations*. 3d ed. New York: Free Press.

Rose, T. 1994. *Black Noise: Rap Music and Black Culture in Contemporary America*. Hanover, N.H.: Wesleyan University Press.

Rosenberg, C. E. 1976. *No Other Gods: On Science and American Social Thought*. Baltimore: Johns Hopkins University Press.

———. 1977. Rationalization and Reality in the Shaping of American Agricultural Research, 1875–1914. *Social Studies of Science* 7: 401–422.

Rossiter, M. W. 1975. *The Emergence of Agricultural Science: Justus Liebig and the Americans, 1840–1880*. New Haven: Yale University Press.

Rudwick, M. J. S. 1976. The Emergence of a Visual Language for Geological Science, 1760–1840. *History of Science* 14: 149–195.

Rudy, A. 2003. The Social Economy of Development. In *Fighting for the Farm: Rural America Transformed*, ed. Jane Adams, 25–46. Philadelphia: University of Pennsylvania Press.

Rudy, A., F. H. Buttel, L. Busch, and W. H. Friedland. 1991. *Toward a New Political Economy of Agriculture*. Boulder, Colo.: Westview Press.

Rudy, A., D. Coppin, J. Konefal, B. T. Shaw, T. Ten Eyck, C. Harris, and L. Busch. 2007. *Universities in the Age of Corporate Science: The UC Berkeley–Novartis Controversy*. Philadelphia: Temple University Press.

Rycroft, R. W. 1991. Environmentalism and Science: Politics and the Pursuit of Knowledge. *Knowledge: Creation, Diffusion, Utilization* 13: 150–169.

Sacks, H., E. A. Schegloff, and G. Jefferson. 1974. A Simplest Systematics for the Organization of Turn-Taking for Conversation. *Language* 50: 696–735.

Sanders, M. E. 1999. *Roots of Reform: Farmers, Workers, and the American State, 1877–1917*. Chicago: University of Chicago Press.

Schaffer, S. 1991. The Eighteenth Brumaire of Bruno Latour. *Studies in History and Philosophy of Science* 22: 174–192.

Schegloff, E. A. 1992. Repair after Next Turn: The Last Structurally Provided Defense of Intersubjectivity in Conversation. *American Journal of Sociology* 97: 1295–1345.

———. 1997. Third Turn Repair. In *Towards a Social Science of Language: Papers in Honor of William Labov*, ed. G. R. Guy, M. C. Feagin, D. Schiffrin, and J. Baugh, 31–40. Amsterdam: John Benjamins.

Schegloff, E. A., G. Jefferson, and H. Sacks. 1977. The Preference for Self-Correction in the Organization of Repair for Conversation. *Language* 53: 361–382.

Scheuring, A. F. 1988. *A Sustaining Comradeship: The Story of University of California Cooperative Extension, 1913–1988*. Berkeley: University of California Division of Agriculture and Natural Resources.

———. 1995. *Science and Service: A History of the Land-Grant University and Agriculture in California*. Oakland: University of California Division of Agriculture and Natural Resources.

Schlosser, E. 2001. *Fast Food Nation*. Boston: Houghton Mifflin.

Scott, J. 2001. *Power*. Cambridge, Mass.: Blackwell.

Scott, J. C. 1976. *The Moral Economy of the Peasant: Rebellion and Subsistence in Southeast Asia*. New Haven: Yale University Press.

————. 1998. *Seeing Like a State: How Certain Schemes to Improve the Human Condition Have Failed.* New Haven: Yale University Press.

Scott, R. V. 1970. *The Reluctant Farmer: The Rise of Agricultural Extension to 1914.* Urbana: University of Illinois Press.

Scruggs, O. M. 1957. *Braceros,* "Wetbacks," and the Farm Labor Problem: Mexican Agricultural Labor in the United States, 1942–1954. Ph.D. diss., Harvard University, Cambridge, Mass.

Shapin, S. 1988a. Following Scientists Around. *Social Studies of Science* 18: 533–550.

————. 1988b. The House of Experiment in Seventeenth-Century England. *Isis* 79: 373–404.

————. 1989. The Invisible Technician. *American Scientist* 77: 554–563.

————. 1994. *A Social History of Truth: Civility and Science in Seventeenth-Century England.* Chicago: University of Chicago Press.

Shapin, S., and S. Schaffer. 1985. *Leviathan and the Air-Pump: Hobbes, Boyle, and the Experimental Life.* Princeton, N.J.: Princeton University Press.

Sims, B. 1999. Concrete Practices: Testing in an Earthquake Engineering Laboratory. *Social Studies of Science* 29: 483–518.

Smith, P. C. 1936. It DID Happen in Salinas. *San Francisco Chronicle,* September 23–26.

Spaargaren, G., and A. P. J. Mol. 1992. Sociology, Environment, and Modernity: Ecological Modernization as a Theory of Social Change. *Society and Natural Resources* 5: 323–344.

Spreckels. 1938. Beet Machinery Program Gets under Way. *Spreckels Sugar Beet Bulletin* 2: 1.

————. 1942a. Advertisement. *Spreckels Honey Dew News,* September, 24.

————. 1942b. Factory 1 Starts and Stops. *Spreckels Honey Dew News,* August, 18.

————. 1943. *Spreckels Sugar Beet Bulletin,* August, 1.

————. 1944. Spreckels Orders Marion Harvesters: 82-Row Harvesters Ordered for 1944 Crop. *Spreckels Sugar Beet Bulletin* 8: 1–2.

————. 1945. Advertisement. *Spreckels Honey Dew News,* February, 24.

————. 1946a. Advertisement. *Spreckels Honey Dew News,* January, 24.

————. 1946b. *Spreckels Sugar Beet Bulletin,* January/February.

————. 1946c. *Spreckels Sugar Beet Bulletin,* July/August.

———. 1951. *Spreckels Sugar Beet Bulletin*, March/April.

———. 1953. *Spreckels Sugar Beet Bulletin*, January/February.

———. 1954. A Grower's Own Story of Complete Mechanical Thinning. *Spreckels Sugar Beet Bulletin* 18: 18–19.

———. 1965. 1965: The Year of the All-Mechanical Sugar Beet. *Spreckels Sugar Beet Bulletin* 29: 38.

Star, S. L. 1995. The Politics of Formal Representations: Wizards, Gurus, and Organizational Complexity. In *Ecologies of Knowledge: Work and Politics in Science and Technology*, ed. S. L. Star, 88–118. Albany: SUNY Press.

Star, S. L., and E. M. Gerson. 1986. The Management and Dynamics of Anomalies in Scientific Work. *Sociological Quarterly* 28: 147–169.

Star, S. L., and J. R. Griesemer. 1989. Institutional Ecology, "Translations," and Boundary Objects: Amateurs and Professionals in Berkeley's Museum of Vertebrate Zoology, 1907–1939. *Social Studies of Science* 19: 387–420.

Starr, K. 1985. *Inventing the Dream: California through the Progressive Era*. New York: Oxford University Press.

———. 1996. *Endangered Dreams: The Great Depression in California*. New York: Oxford University Press.

Steinbeck, J. 1939. *The Grapes of Wrath*. New York: Penguin.

Stoll, S. 1998. *The Fruits of Natural Advantage: Making the Industrial Countryside in California*. Berkeley: University of California Press.

Suchman, L. 1987. *Plans and Situated Actions: The Problem of Human-Machine Communication*. New York: Cambridge University Press.

———. 1996. Constituting Shared Workplaces. In *Cognition and Communication at Work*, ed. Y. Engeström and D. Middleton, 35–60. New York: Cambridge University Press.

———. 2000. Embodied Practices of Engineering Work. *Mind, Culture, and Activity* 7: 4–18.

Sudnow, D. 1978. *Ways of the Hand: The Organization of Improvised Conduct*. Cambridge, Mass.: Harvard University Press.

Tavernetti, A. A. 1939. Thinning Experiments Show Probable Effect of Mechanization of Sugar Beet Thinning. *Spreckels Sugar Beet Bulletin* 2: 1–2.

———. 1943. How Much Have Thinning Costs Increased? *Spreckels Sugar Beet Bulletin* 7: 25–26.

———. 1946. Planting Sugar Beets in Monterey County on 50-Inch Beds. *Spreckels Sugar Beet Bulletin* 10: 4.

———. 1954. Appraisal of the Agricultural Extension Service in Monterey County, California. University of California Agricultural Extension Service, Salinas.

Taylor, P. S., and T. Vasey. 1936a. Historical Background of California Farm Labor. *Rural Sociology* 1: 281–295.

———. 1936b. Contemporary Background of California Farm Labor. *Rural Sociology* 1: 401–419.

Tesh, S. N. 2000. *Uncertain Hazards: Environmental Activists and Scientific Proof.* Ithaca, N.Y.: Cornell University Press.

Thomas, R. J. 1985. *Citizenship, Gender, and Work: Social Organization of Industrial Agriculture.* Berkeley: University of California Press.

Thomas Jefferson on Politics and Government. 2004. Advantages of Agriculture. University of Virginia Alderman Library. <http://etext.lib.virginia.edu/jefferson/quotations/jeff1320.htm>.

Traweek, S. 1988. *Beamtimes and Lifetimes: The World of High Energy Physics.* Cambridge, Mass.: Harvard University Press.

USCCL (U.S. Commission on Country Life). 1911. *Report of the Commission on Country Life.* New York: Sturgis and Walton.

USDA (U.S. Department of Agriculture). 1898. *Progress of the Beet Sugar Industry in the United States.* Washington, D.C.: Government Printing Office.

———. 1904. *Progress of the Beet Sugar Industry in the United States in 1904.* Washington, D.C.: Government Printing Office.

———. 2004a. Crop Production, 2003 Summary. <http://usda.mannlib.cornell.edu/reports/nassr/field/pcp-bban/cropan04.txt>.

———. 2004b. Vegetables, 2003 Summary. <http://usda.mannlib.cornell.edu/reports/nassr/fruit/pvg-bban/vgan0104.txt>.

USEPA (U.S. Environmental Protection Agency). 2003. National Management Measures to Control Nonpoint Source Pollution from Agriculture. Washington, D.C.: Government Printing Office.

U.S. Senate. 1940. Violations of Free Speech and the Rights of Labor: Proceedings of the Committee on Education and Labor, United States Senate, 76th Congress, 3rd Session. Washington, D.C.: Government Printing Office.

Van Loon, J. 2002. *Risk and Technological Culture: Towards a Sociology of Virulence.* New York: Routledge.

Vaught, D. 1999. *Cultivating California: Growers, Specialty Crops, and Labor, 1875–1920*. Baltimore: Johns Hopkins University Press.

Warner, K. D. 2007. *Agroecology in Action: Extending Alternative Agriculture through Social Networks*. Cambridge, Mass.: MIT Press.

Weber, M. 1946. Class, Status, Party. In *From Max Weber: Essays in Sociology*, ed. H. H. Gerth and C. W. Mills, 180–195. New York: Oxford University Press.

Wells, M. J. 1996. *Strawberry Fields: Politics, Class, and Work in California Agriculture*. Ithaca, N.Y.: Cornell University Press.

White, R. 1995. *The Organic Machine*. New York: Hill and Wang.

Wolf, S. 1998. Privatization of Crop Production Information Service Markets. In *Privatization of Information and Agricultural Industrialization*, ed. S. Wolf, 151–182. Boca Raton, Fla.: CRC Press/Lewis Publishing.

———. 2006. Commercial Restructuring of Collective Resources in Agrofood Systems of Innovation. In *The New Political Sociology of Science: Institutions, Networks, and Power*, ed. S. Frickel and K. Moore, 91–121. Madison: University of Wisconsin Press.

Wynne, B. 1989. Sheepfarming after Chernobyl: A Case Study in Communicating Scientific Information. *Environment* 31: 10–15, 33–39.

———. 1996. May the Sheep Safely Graze? A Reflexive View of the Expert-Lay Divide. In *Risk, Environment, and Modernity: Towards a New Ecology*, ed. S. Lash, B. Szerszynski, and B. Wynne, 44–83. Thousand Oaks, Calif.: Sage.

Index

Note: Page numbers in *italics* refer to illustrations or tables.

Inside Technology

edited by Wiebe E. Bijker, W. Bernard Carlson, and Trevor Pinch

Kathryn Henderson, *On Line and On Paper: Visual Representations, Visual Culture, and Computer Graphics in Design Engineering*

Christopher R. Henke, *Cultivating Science, Harvesting Power: Science and Industrial Agriculture in California*

Christine Hine, *Systematics as Cyberscience: Computers, Change, and Continuity in Science*

Anique Hommels, *Unbuilding Cities: Obduracy in Urban Sociotechnical Change*

David Kaiser, editor, *Pedagogy and the Practice of Science: Historical and Contemporary Perspectives*

Peter Keating and Alberto Cambrosio, *Biomedical Platforms: Reproducing the Normal and the Pathological in Late-Twentieth-Century Medicine*

Eda Kranakis, *Constructing a Bridge: An Exploration of Engineering Culture, Design, and Research in Nineteenth-Century France and America*

Christophe Lécuyer, *Making Silicon Valley: Innovation and the Growth of High Tech, 1930–1970*

Pamela E. Mack, *Viewing the Earth: The Social Construction of the Landsat Satellite System*

Donald MacKenzie, *Inventing Accuracy: A Historical Sociology of Nuclear Missile Guidance*

Donald MacKenzie, *Knowing Machines: Essays on Technical Change*

Donald MacKenzie, *Mechanizing Proof: Computing, Risk, and Trust*

Donald MacKenzie, *An Engine, Not a Camera: How Financial Models Shape Markets*

Maggie Mort, *Building the Trident Network: A Study of the Enrollment of People, Knowledge, and Machines*

Peter D. Norton, *Fighting Traffic: The Dawn of the Motor Age in the American City*

Helga Nowotny, *Insatiable Curiosity: Innovation in a Fragile Future*

Nelly Oudshoorn and Trevor Pinch, editors, *How Users Matter: The Co-Construction of Users and Technology*

Shobita Parthasarathy, *Building Genetic Medicine: Breast Cancer, Technology, and the Comparative Politics of Health Care*

Paul Rosen, *Framing Production: Technology, Culture, and Change in the British Bicycle Industry*

Susanne K. Schmidt and Raymund Werle, *Coordinating Technology: Studies in the International Standardization of Telecommunications*

Wesley Shrum, Joel Genuth, and Ivan Chompalov, *Structures of Scientific Collaboration*

Charis Thompson, *Making Parents: The Ontological Choreography of Reproductive Technology*

Dominique Vinck, editor, *Everyday Engineering: An Ethnography of Design and Innovation*